EUREKA!

Also by Michael Macrone

Brush Up Your Shakespeare!
It's Greek to Me!: Brush Up Your Classics
By Jove!: Brush Up Your Mythology
Brush Up Your Bible!

EUREKA!
What Archimedes Really Meant and
80 Other Key Ideas Explained

by Michael Macrone

Illustrations *by* Tom Lulevitch

Cader Books

HarperPerennial
A Division of HarperCollins*Publishers*

The Library of Congress has catalogued the hardcover edition as follows:

Macrone, Michael.
   Eureka! : what Archimedes really meant / Michael Macrone ;
 illustrations by Tom Lulevitch.
     p. cm.
   Includes index.
   "Cader Books."
   ISBN 0-06-270096-0
   1. Philosophy—Miscellanea. 2. Science—Miscellanea. 3. Humanities—Miscellanea.
 I. Title.
 B68.M29   1994
 001—DC20   94-19424

ISBN 0-06-272066-x
96 97 98 99 HC 10 9 8 7 6 5 4 3 2 1

CONTENTS

SCIENCE AND MATHEMATICS

THE HUMAN SCIENCES

INTRODUCTION

Why did Archimedes leap from his tub and run naked through the streets shouting, "Eureka!"? Who was Occam, and what did he shave with his razor? What is a quantum and where does it leap? Do you ever feel the touch of the invisible hand? Does deconstruction really spell an end to Western civilization?

This book answers all these questions and more, taking revolutionary and perplexing ideas of Western thought and extracting their essence. From Greek philosophy to contemporary economics, physics to architecture, *Eureka!* covers some of the most often heard but least understood doctrines and explains them as simply and entertainingly as possible. As Francis Bacon said, "Knowledge itself is power," and the philosophy of *Eureka!* is that knowledge can also be fun.

But what are "great ideas"? Whenever I mentioned writing a book about them, friends asked, "Oh, you mean, like the steam engine?" or "Are you including the guillotine?" The wheel and sliced bread were also noted, along with radioactivity and penicillin and Apollo 8. All these are indeed great, but they're not exactly "ideas"—rather, they're great advances in craft, experiment, and technology.

This book really *is* about ideas, which are more abstract things: theories, laws, principles, paradoxes, aphorisms, constructs, complexes, fallacies, and outrageous claims. It's about bold leaps of thought and carefully constructed doctrines: intellectual (rather than technical) advances in philosophy, science, and culture.

A book this small could hardly cover all the bases, even superficially. But I tried to include those conceptual catchphrases most often cited in popular writings and "cultured" conversation. These comprise principles and laws named after such personalities as Pythagoras, Plato, Newton, Hume, Parkinson, and Oedipus;

provocative statements like "God is dead," "Less is more," "$E = mc^2$," and "The medium is the message"; the mind-bending pardoxes of Zeno, Russell, and Gödel; the big -isms from idealism to utilitarianism to existentialism; and funny foreign phrases such as *cogito ergo sum, tabula rasa,* and *différance.*

I've attempted to arrange this chaos of concepts into logical chapters, beginning with God and ending with a grab bag. The three major sections are Philosophy, Science and Mathematics, and The Human Sciences (comprising economics, psychology, linguistics, and so forth); within each section ideas are arranged in related clusters suitable for continuous reading. Everything you really ever wanted to know about particle physics, for example, may be found in the chapter "From Here to Uncertainty"; you can follow the development of Freud's thought in the chapter "Oedipus, Less Complex."

"Less complex" is the key: Though the depth of a great idea often lies in the details, it's quite possible for anyone who's so inclined to grasp the key issues, the outlines of thought, and the reason an idea is great. It might seem foolish to try covering relativity in eight pages, or evolution in four; but that's ample room to clear up basic misconceptions and offer a modest sketch. Anyone who wants to delve deeper should consult the Appendixes for my list of sources (from which I quote directly) and references (which I used for background).

Beyond the musty library stacks and the dim office, I had many stimulating conversations with Michael Cader, Catherine Karnow, and the assorted wits and experts virtually residing on the WELL computer conferencing system. Their challenges and their support made this a better book. My thanks also go out, as always, to Hugh Van Dusen and Stephanie Gunning at Harper-Collins for their continued support of my series of books on the literature and thought of Western culture.

Eureka!

EUREKA!

PHILOSOPHY

Watchmaker, Gambler, or What?: Ideas of God

Original Sin

For God, the author of natures, not of vices, created man upright; but man, being of his own will corrupted, and justly condemned, begot corrupted and condemned children. For we all were in that one man, since we all were that one man who fell into sin by the woman who was made from him before the sin. To be sure, the particular forms in which we as individuals were to live had not yet been created and distributed; but already the seminal nature was there from which we were to be propagated; and since this was vitiated by sin, bound by the chain of death, and justly condemned, man could not be born in any other state.

St. Augustine, *The City of God* (410), Book XIII

We're starting with a simple one, you'd think. Adam, Eve, fruit, serpent, sin: What more is there to it?

Actually, lots. In the biblical tale of man's first disobedience—eating of the Tree of Knowledge—never once does the Hebrew word for "sin" appear. (It will have to wait for Cain.) Neither does the phrase (or doctrine) "original sin" show up anywhere in the Old or New Testament.

There are hints, certainly. "Of every tree of the garden [of Eden] thou mayest freely eat," says God to Adam (Genesis 2). "But of the tree of the knowledge of good and evil, thou shalt not eat of it: for in the day that thou eatest thereof thou shalt surely die." Since Adam and Eve don't die on that day, God must mean rather what the Hebrew text says: "you shall be doomed to death." And so, therefore, are we all.

But even granting that such a transgression happened, that it was a "sin," and that we owe our mortality to it, we still don't have true "original sin." For believers in the doctrine hold that Adam and Eve's act left an enduring imprint on every human soul. We are not simply going to die, we are *born into* sin—tainted before we ever have a chance.

St. Paul says something along these lines in his Epistle to the Romans: "Wherefore, as by one man sin entered into the world, and death by sin; and so death passed upon all men, for that all have sinned" (Romans 5). You can read this passage any number of ways—as Christians have—for both the word *death* and the cause-effect relationships are ambiguous. Does Paul really mean sin is inborn, or is he speaking only of the spiritual death we invite by sinning?

The basic view, though based in Genesis and Romans, traces more directly to the writings of the Church's greatest classical theologian, St. Augustine of Hippo (354–430). Augustine's account of original sin, in his *City of God,* shares none of Paul's ambiguity. First, Augustine distinguishes death of the body from death of the soul, one of which is inevitable and the other conditional. We are physically mortal, but if we are "freed by grace" our souls may be rescued from perdition.

Second, he insists that both physical death and the state of sin are inborn legacies of Adam's transgression. "For as man the parent is, such is man the offspring." By his own free will, Adam sank into corruption, and this he passed on to all his children. (By the way, it is anything but clear in Genesis that all mankind descended from Adam.)

In formulating his views on sin Augustine was trying to put the lid back on a very large can of worms. One of the most vexing theological issues, then as now, is the problem of evil. If God is all-good, all-knowing, and all-powerful then how can there be evil in the world? How can perfect goodness be the source of badness, or even allow it? This problem doesn't arise in polytheistic religions, as no one god is all-good or all-powerful; and when gods clash only evil can result.

One answer to this riddle was offered by a contemporary of Augustine's, a monk named Pelagius. Evil, Pelagius said, was simply the direct result of human actions, which are freely chosen. If you choose to disobey God's laws, God will visit evil upon you. If enough people act badly, then widespread disaster will follow. Pelagius's theory, except in its stress on free will, is in fact consistent with what we find in the Old Testament: the wind of human sin and the whirlwind of divine vengeance.

The problem with the theory, from the orthodox angle, was that it gave mere mortals the power to screw up the goodness of creation. Man, rather than God, is in the driver's seat. People may *choose* to act well or wickedly, but then God is *compelled* to punish them. By his behavior, man gets to decide whether his soul is saved or damned. What, then, of God's omniscience and omnipotence?

Here's where Augustine came in (along with the Church, which branded Pelagius a heretic). Men cannot "choose" to sin, because sin is not an act but a state of being, which we inherit at birth. Corruption is in our nature, dyed in so deeply that no good deed or string of deeds can wash it away. Because of original sin, mankind is rightly doomed to physical death and eternal suffering. The only way out of the latter is God's merciful grace, which He may grant or withhold as He, and only He, sees fit. So free will, at least when it comes to sin and salvation, is a diabolical illusion.

As if this weren't already depressing enough, it turns out that since God is omniscient, He already knows long before you're born whether or not He's going to save you. This idea of Augustine's is the philosophical basis for Calvinistic theology, which holds that men are divided into two groups, the "elect" (predestined for salvation) and the "preterite" (predestined to hell).

Of course, this brings us back to square one. How could a benevolent God predestine most of the human race to hellfire? Or, to take the problem back to its root, why did God make Adam and Eve capable of sin in the first place? There's really no end to the arguments, which played a leading role in the Protestant schisms of the fifteenth and sixteenth centuries.

Arguments over free will and predestination aren't confined to theology, either. Some materialists—believing that mind is really nothing but matter—argue for determinism on scientific grounds. Since the brain is a physical object, the theory goes, it must obey physical laws, and so thought and behavior, along with everything else in the universe, follow a determined path. At least in theory, given the state of the universe now, everything else is predictable, including what you're going to eat for lunch every day for the rest of your life. Of course, if that's true, then whether you believe it or not is also predestined, which makes arguing the issue sort of pointless.

The Prime Mover

Our present position, then, is this: We have argued that there always was motion and always will be motion throughout all time, and we have explained what is the first principle of this eternal motion: we have explained further which is the primary motion and which is the only motion that can be eternal: and we have pronounced the first movement [or "Prime Mover"] to be unmoved.

Aristotle, *Physics,* Book VIII, chapter 9

You're probably familiar with the basic idea behind Aristotle's "Prime Mover." Everything that happens is caused by something else. Let's say a heavy rain causes a flood in your basement. What caused it to rain? Humidity. But what caused it to get humid enough? Thus the questions go on. Everything that causes something is itself caused by something else, and we can trace the chain of causes as far back as we like. But sooner or later we've got to reach a first cause that just *was,* causing but itself uncaused. This is the Prime Mover.

What prompted Aristotle's thinking was the philosophy of Parmenides, who with the help of his paradoxical disciple Zeno proved that motion is impossible. Parmenides proposed the following: If something exists, it *is,* and it is not what it *is not.* But for this thing to move, it must go from where it *is* to where it *is not.* But then it would no longer be what it *is.* Ergo, motion, or any other kind of change, is impossible, and thus what we perceive as motion and change are illusions.

Not a very compelling argument, but it did cause some trouble at the time. Aristotle hoped to lay Parmenides' argument to rest, and he began by showing that its logic is circular. To say that what exists just *is* is merely a tautology, and it ignores the fact that there are many different sorts of being, which may be divided into qualities and categories that can be combined and separated. Aristotle did agree that at some level what exists is stable and unchanging, for in fact if we are to speak of change or motion we must agree that it is some *thing* that changes or moves. But this level of reality—which he called "matter"—may take on any number of qualities, shapes, and positions, which Aristotle collectively named "forms."

Matter and form are the two essential components of reality, according to Aristotle; matter remains what it is, even as it assumes new forms. (A tree may take on the form of wooden boards and planks, which may take on the form of a chair, which may be painted red, etc.) But this raises the questions of *how* and *why* things move and change. The hows and whys Aristotle called "causes," by which he meant "means of understanding how things come to be." He distinguished four such causes—material, formal, efficient, and final, but I won't go into the fine points here. Suffice it to say that the important sort of cause for Aristotle's purposes is efficient cause, the agency that initiates any change. The efficient cause of a chair, for example, is the carpenter who shapes matter into form.

Now, at first Aristotle believed that motion is an inherent property of matter. Air and fire tend to rise, and water and earth to fall, just because they do; it's built into the material. Likewise the heavens move in a circle because they are made of an element he called "*aether*," which naturally moves in circles. But eventually he came to doubt this theory, partly because it's pretty circular itself, and partly because it doesn't account very well for natural phenomena.

This is where the idea of efficient cause becomes important. It's not enough to say that things "naturally" move up, down, or around, because that avoids the issue of who or what made them move that way to begin with. The same question arises about the motions of animate beings. Say I move myself to get a beer from the refrigerator. What causes me to do it? I'm thirsty: That's the efficient cause. But what caused me to be thirsty? We could follow such a line of questioning from cause to cause to cause, but at the pain of going on forever we must assume that at some point we will reach a cause that is not itself caused by anything else.

Aristotle claimed that every change or motion ultimately traces back to the same, single "uncaused cause" and "unmoved mover," which he named the "Prime Mover."

What's more, he asserted that all things move and change in order to approach some goal or "final cause." For all things this goal is perfection: everything strives to be all that it *can* be. This final cause, this perfection, is one and the same with the Prime Mover, whose perfection is expressed by the fact that it neither changes nor moves.

From a scientific angle, Aristotle understood the Prime Mover (PM) as the end of the causal chain, an immaterial and unchanging principle that sets all other things in motion, directly or indirectly. (In fact, the only thing the PM moves directly is the outermost heavens.) Being a philosopher, Aristotle also had a metaphysical view of the PM: Being perfect, it must be the same as "thought" (philosophy).

Finally, Aristotle took the PM theologically. As the source of animation, the PM must itself be animate; as thought, it must constantly think. If it thought about the shifting, imperfect things of this world, its thoughts, following their objects, would also be changeable and imperfect. But as the PM is perfection itself, this is impossible. Therefore the PM is thought that thinks about itself, perfection that considers its own perfection. Who else could this be but God?

Good point, thought the thirteenth-century theologian St. Thomas Aquinas, who used the same argument to prove the existence of an "unmoved mover," who is God. But Aquinas, like Aristotle, relies on quite a number of unprovable assumptions— for example, that all motions and causes must logically trace back to a single primary entity. It would be just as logical to argue that causes travel in circles, or that causes are determined purely by physical laws, or that behind each event there is a multiplicity of causes that themselves have many causes, and so on, producing an infinity of "original" causes rather than just one. You could even argue—as indeed some have—that causality is a fiction, a creation of the human mind. But then nobody's ever had much luck proving the existence of God on logical grounds, as we shall continue to see in this chapter.

Occam's Razor

Plurality is not to be assumed without necessity.

William of Ockham, *Quodlibeta*, Book v (ca. 1324)

William of Ockham ("Occam" is the Latin spelling), an English theologian of the early fourteenth century, is at best obscure today. Thomas Aquinas and Duns Scotus are superstars by comparison, and yet it is Occam whose thought prefigured modernity.

The one thing some *do* remember is Occam's so-called "razor," the logical implement he wielded to trim absurdities out of arguments. Occam's maxim was that the simpler an explanation is, the better. If it isn't *necessary* to introduce complexities or hypotheticals into an argument, don't do it; not only will the result be less elegant and convincing, it will also less likely be correct.

As we shall see, one hypothetical Occam's razor dispensed with was the existence of God. Not that he didn't believe God exists, of course; he just thought you couldn't prove it, because to do so you had to resort to rather complex (and incredible) arguments. Theologians wanted a *scientific* proof of God; but what Occam said, and most everyone eventually accepted, is that science and theology have different objects and require different methods.

Actually, Occam wasn't the first to employ the razor; and nowhere in his work do we find its familiar formulation, "Entities are not to be multiplied without necessity." But he used it with a vengeance, mostly in reaction to the prevailing methods of theology and philosophy. His predecessor Aquinas and other "Scholastics"—a name they earned by preferring texts to experience—dearly wished to make theology scientific. They hoped to resolve apparent contradictions between ancient science and scriptural teaching, and offer rational explanations or proofs of theological concepts (such as God's existence).

One step in this process was to treat universal concepts such as "good" or "great" (and even such mundane universals as "tree" or "dog") as real, independent entities. If we call both this elm and that oak "trees," then there must be some real, existing thing ("treeness") they share. Likewise, if Socrates and Parmenides are both good, it is because there is such a thing as goodness that they both have. Such a doctrine—which is rather more Platonic than Aristotelian—is known as "realism."

Occam thought realism was just so much nonsense, a confusion of categories elevated to a science. It is a mistake, he thought, to treat names as *realities* rather than as *descriptions*. (The idea that names are just names is called "nominalism.") If we call both this elm and that oak "trees," it is because we have decided what makes a tree a tree, not because "treeness" exists separately in reality. If all trees were suddenly to disappear, there wouldn't be any "treeness" left to talk about, except as a memory or pure abstraction.

Occam used the razor to do away with realist universals, insisting that valid explanations must be based on simple and observable facts, supplemented by pure logic. Accepting these conditions means that we won't be able to scientifically prove God's existence or His goodness, or any of the other tenets of faith. Such a conclusion didn't bother him at all; he thought theology was one thing (a matter of revelation) and science another (a matter of discovery). This idea took a while to prevail, as Galileo could have told you, but ultimately science and religion went their own separate ways. In great part, this is what modernism is all about.

The Ontological Proof

The "ontological proof" was yet another attempt to demonstrate beyond doubt that God exists. Devised by St. Anselm of Canterbury (1033–1109), an Italian, it goes something like this: So long as we can imagine absolute perfection, then it must exist. If it exists, it is God.

Not convinced? Let's look more closely at the proof, dubbed "ontological" by Immanuel Kant after the Greek word for "being" (*ontos*). Anselm began with the following experiment: Imagine a being more perfect than any other you can conceive. If you understand that sentence, you must have some concept of such a being; otherwise the sentence would be incomprehensible. (Just as the sentence "Imagine a unicorn" only makes sense if you have some concept of "a unicorn.")

Now, is this being you imagine (call it "B") simply a fantasy? Anselm thought not. For if B does not exist, then you could imagine a more perfect being, namely a being just like B but *who also exists.* For obviously a real good is more perfect than an imaginary good. Thus the assumption that B is a fantasy must be false, for if it were true we could imagine a more perfect being, which contradicts the hypothesis.

So B exists, and Anselm defines it as God. In other words, God is precisely that being we take to be the most perfect imaginable. If we had merely sought after the most perfect existing being, we would not have arrived at Anselm's conclusion, for we wouldn't have been able to prove that what we found is God. The key to his proof lies in the *concept* of being as a kind of perfection in itself, not in any prior experience of its existence.

But that's exactly the problem with Anselm's ontological proof. By building existence into the definition of "the most perfect conceivable being," then to say that such a being exists is merely to restate its definition. Otherwise, contradicting it would

not necessarily be false, as the proof requires. Existence and perfection refer to exactly the same thing. Anselm's proof is thus, to use Kant's words, merely "a miserable tautology."

Kant wasn't the first to realize that Anselm's demonstration had problems. In fact, one of Anselm's contemporaries, Gaunilo of Marmoutier, pointed out that the ontological proof could be used to prove the existence of almost anything. Gaunilo's specific example was that of the perfect island, better than any known island, a place of every conceivable delight. Since we can imagine such a thing, we must have a concept of it; and if it didn't exist, we could conceive of a more perfect (that is, existing) island; therefore, it must exist.

As Anselm himself responded, Gaunilo misses the point. For the concept of an island does not involve the concept of existence, just as the concept of a perfect circle does not depend on the existence of any such circle. The concept of a *being,* however, necessarily involves the concept of existence. We can easily imagine that a perfect island or a perfect circle does not exist; but we cannot imagine that the most perfect conceivable does not exist, because the very concept precludes it. What may possibly not exist is by definition lesser than what cannot *not* exist. This logic sold Descartes, Spinoza, and Leibniz, among the big guns of philosophy, on the validity of Anselm's proof.

It would be more than seven centuries before Kant finally disposed of the proof. In his *Critique of Pure Reason* (1781), he showed that Anselm had mixed up his categories, treating a grammatical unit (the predicate "to be") as an ontological quantity. To say that a thing "is" or "exists" is not, according to Kant, to add anything to it. Rather, it is to state that something in reality corresponds to a concept we have. To say that "this chair exists" is not to add anything to the chair, but only to make a judgment about it—that our experience shows it to be real. We may only say that something "is" or "exists" if we can experience it; the truth of such a statement depends on a correspondence between word or concept and a thing in reality.

In short, if God does not exist, he cannot be "improved" or made more perfect by adding existence to him, since there is no "Him" to add anything to. If the predicate "exists" disappears, so does the subject "God" (or "most perfect conceivable being," or "chair," or whatever subject we have proposed). Likewise, to say "God does not exist" doesn't "subtract" anything from God, since we're only proposing that there is no such being "God" to take anything from—in this case, "God" is a grammatical, not an actual, subject.

In other words, there is no logical contradiction in the statement that "the most perfect conceivable being does not exist": we are stating, or attempting to state, that such a being has no objective reality, rather than contradicting the notion of such a being in itself. And if the negative ("X does not exist") is not logically contradictory, then the positive ("X exists") is not logically necessary. The only true test of whether something exists is experience.

And that was effectively the end of Anselm's "proof," although there have since been many attempts to salvage some such argument from the ruins. None has proved successful, insofar as all of them involve some confusion of categories, but you've got to admire people for trying.

Pascal's Wager

Let us consider the point and say: "Either God exists, or he does not exist." But which of the alternatives shall we choose? Reason can determine nothing: there is an infinite chaos which divides us. A coin is being spun at the extreme point of this infinite distance which will turn up heads or tails. What is your bet?

Blaise Pascal, *Pensées* (posthumous edition, 1844)

God may not "play dice" [see p. 105], but we all play dice with God. That was the conclusion of the seventeenth-century French mathematician Blaise Pascal as he took up the bedeviling question of God's existence.

Pascal, unlike Anselm, conceded that it is impossible to "prove" that God exists—in fact, he claimed, human reason is incapable of proving anything for certain. The important question for him was whether one *ought* to believe God exists, and his answer was that you would be foolish not to. Pascal's demonstration employs the mathematics of probability, which he helped invent. (He hoped to appeal especially to his aristocratic friends, who were passionate gamblers.)

In Pascal's view, your belief or disbelief in God amounts to a wager. If God does exist and Holy Scripture is true, belief will get you infinite happiness after death. If God does not exist, all you have to lose by believing in Him are the finite pleasures of a finite life. Even if you think the odds on God's existence are near to zero—Pascal suggests they're closer to 50 percent—the only rational thing to do is play the game. (In mathematical terms, any finite percentage of infinity is still infinity.) Ergo, reason dictates that you *must* believe in God.

Of course, you might still resist reason, but that would only be by letting your passions get the better of you. According to Pascal, desires can be tamed by behaving *as if* you believed in God and by participating in good Christian rituals. Once you're used to it, you'll discover that in shedding your filthy habits you're even happier than before—and this, in Pascal's view, is the true payoff of the wager.

Pascal's argument is pretty neat, but as he himself probably knew, multiplying and dividing infinities is a tricky business. Pascal's logic would make pursuing any promise of infinite happiness, religious or otherwise, the rational thing to do if there were a nonzero chance of success. (Say there's a 1 percent chance the Fountain of Youth exists; you should drop everything now and go look for it.)

For Pascal's Wager to work, you have to grant much of what he wants to prove—that if God exists He is infinite, omniscient, omnipotent, and the true author of the Bible. But of course there is an infinite number of other possibilities—for example, that God exists but does not really care about personal behavior or (more damaging to Pascal's argument) God exists but is not in fact an infinite being.

In any case, it is considerably more difficult to act on beliefs you don't hold than Pascal is willing to admit. (And presumably God would know whether you were sincere or just gambling.) As far as human nature goes, certain pleasures usually win out over uncertain ones, no matter how promising. In the heat of passion, infinite possibilities can seem pretty infinitesimal.

"God Is Dead"

Have you heard of that madman who lit a lantern in the bright morning hours, ran to the market place, and cried incessantly, "I seek God! I seek God!" As many of those who do not believe in God were standing around just then, he provoked much laughter....

"Whither is God," he cried. "I shall tell you. *We have killed him*—you and I. All of us are murderers.... God is dead. God remains dead. And we have killed him...."

Friedrich Nietzsche, *The Gay Science* (1882), section 125

Shakespeare did not say "To be, or not to be." He *wrote* it, but Hamlet says it. Neither did Friedrich Nietzsche say "God is dead"; a "madman" does. While it's true that Nietzsche himself went mad at 45, there's still a difference between life and literature, even when the latter is called philosophy.

So what does the madman mean? Not that there are "unbelievers" in the world, for that was always true; nor simply that God does not exist. For if "God is dead," then He must have once been alive; but this is paradoxical, since if God were ever alive, He, being eternal, could never die.

So the madman speaks not of the believer's God, who always was and always will be, but rather of what God represented and meant to his culture. This God was *a shared belief* in God, and it is such belief that, in nineteenth-century Europe, was expiring. Where once God stood—at the center of knowledge and meaning—there is now a void. Science and philosophy alike treat God as irrelevant, and once again man has become the measure of all things.

We Westerners have, in turning ever more toward nature and away from the supernatural, "killed" the God of our ancestors. The unbelievers in Nietzsche's tale think seeking God is rather funny; only the madman realizes the terrible gravity of God's death. Not that he laments it; in fact, he calls it a "great deed," but a deed likely too great for us, the murderers, to bear. "Must not we ourselves become gods simply to seem worthy of it?"

This is the question posed by Nietzsche's parable, which, to return to our first point, is a fiction and not a philosophical statement. Nietzsche in fact abhorred metaphysical speculations on the intelligibility, nature, and existence (or nonexistence) of supernatural abstractions like "God." He couldn't give a hoot for God, but he did have a lot to say about religion—Christianity in particular. To him, religion, by focusing on eternal life, is actually a kind of *death*: It turns us away from life and truth, which are in the world and not in some supernatural never-never land.

What's more, a religion such as Christianity, despite the teachings of Jesus, perpetuates intolerance and conformity, which Nietzsche found especially repugnant. Whatever is old, habitual, normative, or dogmatic, he thought, is contrary to life and to dignity; it manifests what he called a "slave mentality." In a sense, for a man or woman to live, he or she *must* "kill" God—must overcome dogma, conformity, superstition, and fear. This is a necessary first step on the way to becoming, not a god, but "overman" [*see* p. 56].

Ideas from Long Ago:
Greek Philosophy

"Everything Changes but Change Itself"

Everything flows and nothing abides; everything gives way and nothing stays fixed.... You cannot step twice into the same river, for other waters and yet others go ever flowing on.... It is in changing that things find repose....

Heraclitus, fragments

Fellow Greeks knew the philosopher Heraclitus as "the Obscure One," and they had their reasons. Heraclitus (late sixth century B.C.) was perhaps the most delphic of pre-Socratic thinkers. A moody man with a somewhat grim outlook on life, he essentially argued that all things, good and bad, must pass.

Like Thales of Miletus (inventor of Greek philosophy), Heraclitus thought that all things were made from a single, permanent substance, which had to be one of the four "elements"—earth, air, fire, water. Thales chose water; Heraclitus took fire. "The thunderbolt pilots all things," was his cryptic maxim.

The world, he thought, is like the flame of a candle: ever the same in appearance, but ever changing in substance. Ironically, his more famous example of this form/substance paradox is watery: "You cannot step twice into the same river." Though a river might appear to always be the "same" river, its waters flow by ceaselessly. The moment you step into the water, it is gone.

All the world is likewise in constant flux; change is constant and inescapable. To Heraclitus's point that everything changes, others would add the logical conclusion, "but change itself." What he didn't mean is that all is chaos; behind the flux and strife he saw a guiding principle, an organizing force, which he called *logos,* a Greek word meaning "reason" or "logic."

Eureka!

It is this disembodied *logos,* inherent in the universe, that shapes conflict and change into beauty and pleasure. "Opposition brings concord" is one of Heraclitus's paradoxes. "Out of discord comes the fairest harmony." Good does not exist apart from evil, health from disease, satisfaction from hunger, or rest from weariness: They are two sides of the same metaphysical coin, each yielding the other as change turns the coin round and round.

Heraclitus's ideas surfaced again, though slightly changed, in the philosophy of Empedocles (fifth century B.C.), who inspired the Latin poet Horace, four centuries later, to coin the phrase *concordia discors*—"discordant harmony." Plato's notions of material transience and ideal permanence [*see* PLATO'S CAVE, p. 23] also owe a debt to Heraclitus, as (more indirectly) do the sentiments of the Biblical wise man Ecclesiastes, whose song-worthy line, "To everything there is a season," is just one of his many Greek touches.

"Man Is the Measure of All Things"

SOCRATES: Can I ever fail of knowing that which I perceive?

THEAETETUS: You cannot.

SOCRATES: Then you were quite right in affirming that knowledge is only perception; and the meaning turns out to be the same, whether with Homer and Heraclitus, and all that company, you say that all is motion and flux, or with the great sage Protagoras, that man is the measure of all things; or with Theaetetus, that, granting these premises, perception is knowledge.

Plato, *Theaetetus*

If you're familiar at all with Socrates' technique, you will already have guessed that he's setting Theaetetus up for a fall. The idea that "man is the measure of all things" strikes Socrates as vain as well as false; but rather than just saying so outright he gently steers his young interlocutor by way of the "Socratic method" so that Theaetetus will come to understand why it is false. By the end, neither of them has determined the truth, but at least they agree that Protagoras was wrong.

Protagoras (fifth century B.C.) was the founder of a group known as the Sophists, who believed that wisdom could be taught (for a price)—a radical idea at the time. The founding principle of Protagoras's philosophy is that "man is the measure of all things"; in other words, things exist by virtue of how we perceive them. The object world is measured against man, and there is nothing outside man that determines being or truth. This rather abstract idea, which is anathema to Socrates' notion of ideals, has surprisingly become a popular catchphrase. But what we seem to mean by it today is that "our needs and desires determine what counts in this world."

Eureka!

Zeno's Paradox

You've probably heard a version of Zeno's Paradox—or rather, one of Zeno's paradoxes; the Greek philosopher (fifth century B.C.) had a bagful. In fact Zeno, a philosophical gadfly, made paradox his philosophy.

Only one, though, is called "Zeno's Paradox," and it comes in a variety of forms. Today the most common is this: Suppose you're traveling from point A to point B. To reach B, you must first travel half the distance. Once you've arrived at the mid-point, you must then travel half the remaining distance. But once you've arrived at the mid-point of the remaining distance, you still have to travel half the remaining distance.

In fact, this series goes on *ad infinitum*. Since it takes some time, no matter how small, to travel half of any given distance, and since the remaining distance can always be divided in half, it will therefore take you an infinite amount of time to travel from A to B. In short, it is *impossible* to ever reach B.

A more colorful version of the paradox involves a race between the Greek hero Achilles and a lowly tortoise. Let's say Achilles gives the tortoise a head-start; you can prove, using Zeno's logic, that he can't possibly win the race. Assume that Achilles begins running at 1:00. To catch up, he must first reach the place where the tortoise was at 1:00, but that may take 10 minutes. In that ten minutes, the tortoise has traveled a bit farther, so Achilles, to catch up, must now reach the point where the tortoise was at 1:10. This will take some time, say five minutes. But in that five minutes, the tortoise has lumbered farther along toward the finish line, and now Achilles must race to where the tortoise was at 1:15. And so on. Therefore, the tortoise will always be ahead of Achilles, no matter how much faster Achilles may run.

Zeno of course knew that in reality, as it was commonly understood, Achilles or any other healthy person could easily beat the tortoise. He simply didn't believe that the common understanding of reality was coherent, since, as he attempted to demonstrate, common sense and the laws of motion couldn't both be true at once. (Zeno's mistake was that he didn't realize he was dividing infinity by infinity, but you don't really need to know the details.) What Zeno was really trying to prove was the doctrine of his mentor Parmenides, whose notions of being and not-being were rather abstract—according to Parmenides, in fact, reality was unreal.

The Greek philosophers of Parmenides' day couldn't poke effective holes in his arguments; the first to do so was Plato, who attacked Parmenides' doctrines in a series of dialogues (*Parmenides, Theaetetus,* and *Sophist*). But Plato wasn't completely victorious, since Aristotle still thought it necessary to challenge and refute Parmenides' and Zeno's arguments, which he did in the course of examining the causes of motion. Aristotle's ultimate thesis, however, had problems of its own [*see* THE PRIME MOVER, p. 5].

Plato's Cave (Idealism)

"And now," [Socrates] said, "let me show in a figure how far our nature is enlightened or unenlightened: Behold! human beings living in an underground den, which has a mouth open towards the light and reaching all along the den; here they have been from their childhood, and have their legs and necks chained so that they cannot move, and can only see before them, being prevented by the chains from turning round their heads. Above and behind them a fire is blazing at a distance, and between the fire and the prisoners there is a raised way; and you will see, if you look, a low wall built along the way, like the screen which marionette players have in front of them, over which they show the puppets.... [They are like ourselves,] and they see only their own shadows, or the shadows of one another, which the fire throws on the opposite wall of the cave."

Plato, *Republic,* Book 7

Plato (ca. 428–348 B.C.) did not think this the best of all possible worlds. It is a kind of prison, he wrote, where we are locked in darkness and shadows. But beyond this prison lies a bright and hopeful world of truths he called ideas or ideals, which is why we call his doctrine "idealism."

Plato develops his idealistic ideas most memorably in the *Republic,* where as usual his spokesman is his mentor, Socrates. (To what degree Socrates actually held Plato's views is unknown.) Socrates compares the everyday world to an "underground den" or cave where we are chained in place. Before us lies a wall and behind us a fire; unable to turn our heads, we see only the shadows cast on the wall by the fire. Knowing nothing else, we naturally take these shadows for "reality"; our fellow human beings and all the objects in the cave have no reality for us other than this.

But if we could free ourselves from our chains, if we might even just turn toward the mouth of the cave, we would eventually realize our error. At first direct light would be painful and disorienting. But soon we would adapt and begin to perceive the real persons and objects we once knew only in shadow form.

Even so, we would cling out of habit to the shadows, still believing them real and their sources only illusions. But if we were dragged out of the cave and into the light, then sooner or later we would come around to the correct view of things and pity our former ignorance.

Plato's analogy is an attack on our habits of thought. We are, he says, used to accepting the concrete objects around us as "real." But they are not. Or rather, they are only imperfect and less "real" copies of unchanging and eternal "forms." These forms, as Plato defines them, are the permanent, ideal, and original realities from which (somehow) imperfect and corruptible concrete copies are struck off. For example, every chair in our familiar object-world is merely an imitation or "shadow" of the Ideal Chair. Every desk is a copy of the Ideal Desk, which never changes, which exists in eternity, and on which you can never spill coffee.

These ideal chairs and desks, according to Plato, aren't fantasies; they are in fact more "real" than their worldly imitations, because they are more perfect and universal. Yet because our corrupt senses have always been trapped, we are blind to the world of ideals. Our minds are chained to imitations, which we thus mistake for reality. We are prisoners in a philosophical cave.

But all is not lost, for while everywhere man is in chains, philosophy can set us free. If we only let it, it will drag us from the cave of gloom and ignorance into the light of true being. We may for a time object to what we then see, clinging to the "reality" of objects and denying the truth of philosophical Ideals. But sooner or later we will begin to see clearly, and even approach the master idea, the ideal of ideals, which is the Idea of Goodness itself. Of course, being a philosopher, Plato defines Goodness as knowledge.

The Three Laws of Thought

Now since it is impossible that contradictories should be at the same time true of the same thing, obviously contraries also cannot belong at the same time to the same thing.... If, then, it is impossible to affirm and deny truly at the same time, it is also impossible that contraries should belong to a subject at the same time, unless both belong to it in particular relations, or one in a particular relation and one without qualification. But on the other hand there cannot be an intermediate between contradictories, but of one subject we must either affirm or deny any one predicate.

Aristotle, *Metaphysics,* Book IV, chapters 6–7

For more than two millennia, Western logic hewed to three basic "laws of thought." Indubitable on their face, these axioms practically define the way we think. But they are far more complex and less obvious than they seem.

The three laws, as codified by Aristotle, are these:

1) A thing is identical to itself. The standard symbolic expression of this law, called the "law of identity," is "A = A." For example: "Socrates is Socrates."

2) A thing cannot at once both *be* and *not-be*—"A and not-A is false." This is called the "law of contradiction." For example: "It is false that Socrates is at once a man and not-a-man."

3) Given a definite state or quality A, a thing must either have it or not—"Either A or not-A." This is called the "law of the excluded middle," since there is no middle ground between A and not-A. For example: "Socrates is either alive or not-alive."

It's pretty hard to argue with any of these laws, which we normally take completely for granted. But philosophers and mathematicians don't care about what's normal; they care about what's true. Are these laws necessarily true in every possible situation? For about a century now, the answer has been "no."

The doubts began setting in once philosophers thought harder about the meanings of "is" and "not" in Aristotle's laws.

Because such words may be used in a variety of ways, the laws easily fall into semantic muddles. The most interesting of such problems bedevils the law of the excluded middle (3). Take a simple example such as "a rose is either red or it is not." How simple is it really? Perhaps you and I wouldn't agree on how red a rose has to be to be "red"; perhaps we don't even agree on what "red" means. I once had a Volkswagen Rabbit that I called "red" and other people called "orange." We all agreed it was very ugly, but we couldn't agree on how to describe the ugliness.

Qualities ("predicates," to use the logical term) are often subjective. I may think John is tall, but you may not. We might both be right; what is "right"? Can we say that either statement is "true"? Take another example: the statement "Unicorns are vicious." This is false, because unicorns don't exist. But the contrary statement "Unicorns are not vicious" is equally false, for the same reason.

Mathematicians have recently made similar objections to all the laws, or at least to the claim that they're a sufficient basis for logic. It's fine to say that "5 is 5," or that "5 cannot be not-5," or that "5 must either be even or odd." But once we enter the realm of infinite numbers, such statements become meaningless; we cannot prove that an infinite number is *either* even or odd. ("Zero is even" and "1 is prime" are other undecidable propositions.) Similarly, to take an example from physics, we can't say that "light is either a wave or not-a-wave." For reasons such as these the laws of thought have fallen into disfavor, at least in scientific circles.

The laws have also had their problems among philosophers, chief among them G.W.F. Hegel. As far as Hegel was concerned, a thing *can* in some sense be its own opposite. For more on this brain-twisting notion, see THE DIALECTIC, p. 51.

Cogito Ergo Huh?: Beginnings of Modern Philosophy

"I Think, Therefore I Am" *(Cogito Ergo Sum)*

I noticed that while I was trying to think everything false, it must needs be that I, who was thinking this, was something. And observing that this truth, *I think, therefore I am* [*cogito ergo sum*] was so solid and secure that the most extravagant suppositions of the skeptics could not overthrow it, I judged that I need not scruple to accept it as the first principle of philosophy that I was seeking.

René Descartes, *Discourse on Method* (1637)

It may not seem like a major achievement in hindsight, but when René Descartes proved his own existence, it was a very big deal.

One of the simplest, most elegant, and best known philosophical proofs, "I think, therefore I am" struck at the skepticism so fashionable in Descartes's day. Friends and colleagues of the French mathematician/philosopher (1596–1650) held the view that nothing was certain, only probable at best, because the mind is so easily fooled.

But having first made his mark by devising the Cartesian coordinate system, Descartes had a vested interest in proving that at least *some* things (such as mathematical theorems) are indeed absolutely true. Without some basis for certainty, he believed, there could be no true knowledge at all—mere probability just wouldn't do.

Descartes took it for granted that knowledge must ultimately be based on one indisputable fact (rather than, say, two or ten). To find it, he began by accepting the familiar skeptical arguments. Let's assume, along with Descartes, that our senses are unreliable and often lead us to erroneous conclusions. (For example, we see the sun "rise," but it's the earth that's moving; we sometimes mistake dreams and mirages for reality.) Therefore, since they are uncertain and quite possibly illusory, we must discard sense impressions as a basis for knowledge.

What's left is reason, which Descartes and other "rationalists" believed is a more reliable guide than experience. But suppose that reason, too, is fallible, and that even logic may err. Perhaps such seemingly self-evident notions as "$2 + 2 = 4$" and "happiness is good" are actually false, and are planted in our minds by an all-powerful evil demon bent on total deception. Suppose furthermore that the entire world and everything in it, including rational thought, are a dream in this evil demon's mind. We have no way to prove that this is true *or* false; so what could possibly be left as an absolute truth?

Descartes's answer: the very fact that he thought it all up in the first place. That is, no matter what scenario of reality you come up with, you're still *thinking* it. And if you're thinking it, *you* must exist. Or—in the Latin version found in *Discourse on Method*—*cogito ergo sum*: "I think, therefore I am."

Here at last, Descartes concluded, was the foundation of true knowledge—thought itself, and whatever particular thoughts (such as "substance," "self," and "God") are innate in the mind. But here we must pause to ask Descartes a few questions. Let's grant that *cogito ergo sum* is right and ignore the possibility that thought itself may be a mirage. Even so, it isn't an immediate step to innate and certain ideas such as "God"—the content, if not the fact, of our thought could still be a demonic deception.

But for Descartes such conclusions were unimaginable. He was sure he existed, that he thought, and that certain clear and distinct impressions are the essence of thought: Without them, thought cannot happen. And since an omnipotent and benevolent Creator—God—was among those impressions, God must exist. And, being omnipotent and benevolent, God precludes the existence of an omnipotent, deceptive evil demon. Once we dispose of the demon, we dispose of our doubts about logical and mathematical truths.

This still leaves us with the problem of uncertain sense impressions. Descartes thought that God would never allow us to be totally deceived, so we can trust at least that the world exists and that its substance is real. On the other hand, physical substance is absolutely different and separate from thought, which leads to Descartes's famous "mind/body problem."

If the mind exists, where is it? If it is somewhere, it has a physical location and reality, and so it must be some sort of substance. If it has no substance, in what way does it exist? Descartes was unable to solve this riddle (he weakly suggested that the mind is located in the center of the brain—the pineal gland). In fact, nobody's ever come up with a really plausible solution. The trend in science these days is toward defining mind as only a bunch of neurochemical reactions; but I wouldn't hold my breath waiting for proof.

Descartes's rationalism, an outgrowth of the medieval view that truth requires certainty, eventually fell before the gains of empirical science, to which truths are always hypothetical, provisional, subject to improvement, and as dependent on trial and error as on reason. Nonetheless, *cogito ergo sum* is still a great idea—a culmination of ancient philosophy and a stimulus to modern logic and metaphysics. Descartes's coordinate system may have lasted longer, but it is mostly by the *cogito* that we remember him.

Hume's Fork

Hume's fork is a philosophical utensil for separating interesting problems from bogus ones. Proposed by Scotsman David Hume (1711–1776), the basic idea is that every statement or claim falls into one of three categories: 1) either true or false by definition, 2) dependent on experience, or 3) just nonsense. These are the fork's three tines. As far as Hume was concerned, only type 2 statements are interesting, which is why he's called an "empiricist" (from the Latin for "experience").

Actually, though, Hume stole the basic idea from the so-called "rationalist" Gottfried Leibniz (1646–1716), who followed Descartes in believing that reason is a more certain guide to truth than experience is. Leibniz had his own fork, but it only had two prongs: 1) logically necessary assertions and 2) "contingent" (unnecessary) assertions. Examples of necessary assertions are "$2+2=4$" and "A cocker spaniel is a dog." That they are necessary means their negations must be false. Examples of contingent assertions are "Caesar crossed the Rubicon" and "Bill Clinton is president of the United States." Such statements may in fact be true, but their opposites aren't *necessarily* false. Truth or falsity in such cases is not logical, but depends on historical events that may have happened differently.

But Leibniz made this distinction only in order to destroy it. What he believed was that contingent propositions *are* necessary if you look at them the right way. Bill Clinton *had* to win the election in 1992, because the outcome was predetermined by God. There is no possible "alternate reality" in which he could have lost. Everything is as it was meant to be, ergo there are no contingent assertions.

Hume, with his fellow British empiricists, scoffed at such Continental follies. Like Leibniz, Hume distinguished necessary propositions (what he called "relations of ideas") from contingent ones ("matters of fact"), but he insisted that they are indeed distinct. Not only are matters of fact not necessary, but necessary propositions are all but useless. As Hume saw it, any statement that is necessarily true (or "analytic," to use the term Kant coined) is just a tautology: it is empty and tells us nothing. To say that "a cocker spaniel is a dog" is merely to restate the definition of "cocker spaniel." To say that "2 + 2 = 4" is to say nothing new, but just to pursue the consequences of how we have defined the terms.

The only important kind of statement, for Hume, is one about matters of fact, which are not necessarily true and which thus tell us something new about the world. To say that "Julius Caesar crossed the Rubicon" is informative because Julius Caesar might *not* have crossed the Rubicon. Any real knowledge will come in such a form, which Kant called a "synthetic" statement, and such statements are based on observation rather than on reason. This is the essence of empiricism.

Hume realized that there are all sorts of statements that are neither tautologies (analytic) nor informative (synthetic). For example, "Unicorns are vicious" is hardly a logical truth, but on the other hand it tells us nothing about the world, since unicorns don't exist. This kind of statement Hume called "nonsense." As he saw it, most books of theology or metaphysics were full of just the sort of nonsense that is neither necessarily true nor illuminating of facts, and his prescription for such a book was to "commit it then to the flames, for it can contain nothing but sophistry and illusion" (*An Enquiry Concerning Human Understanding,* 1748).

Using the fork might not seem like a big deal now, but at the time Hume's notions had some radical consequences. For example, he showed that the statement "God exists" is neither necessarily true (since its denial isn't necessarily false) nor empirical (since we don't experience God with our senses). By "Hume's fork," the statement must therefore be nonsense—that is, beyond the bounds of knowledge.

Furthermore, Hume's empiricism reduces all experience to the contingent—that is, to things that may or may not be true and that are therefore not determined. Among such things is the experience of cause. We might see a cue ball hit an eight ball, and then see the eight ball roll toward the pocket, but that one thing "causes" another is just an abstract idea, our way of explaining a sequence of events. His conclusions proved very troublesome to science (*see* the next entry).

Hume attempted to replace the empty certainties of mathematics and science with a more meaningful model of reality, based on human psychology, probability, and habitual behavior. For this effort Hume is now called a skeptic, which is true so far as it goes. But he might also be called a psychologist or statistician. In our age of quantum physics and opinion polls, Hume's ideas are virtually taken for granted.

The Scandal of Induction

The scientific method pioneered in the seventeenth century by Francis Bacon, René Descartes, and others is essentially an inductive process. As opposed to deduction (which derives new truths from established ones), induction passes from particular observations to general conclusions.

Bacon and his followers insisted that scientific knowledge can never rest on given truths, whether mathematical or metaphysical, but must ground itself in observation and experiment. True scientists regard the natural world, seek out its patterns, propose hypotheses, and then test them by experiment. A hypothesis becomes theory if established through repeated experiment but is rejected if contradicted ("falsified") by experiment.

Philosopher David Hume, while agreeing that knowledge can only be gained through experience, thought the scientific method had big problems. Exactly what does it mean, Hume asked, to "know" something by observation or experiment—that is, by induction? All you know is that whenever X appears to happen, Y also appears to happen.

Let's say X happens a hundred times, and every single time Y also happens. You throw a rock at a window, and it breaks. You throw 99 rocks through 99 other windows, and they all break, too. You conclude by induction that throwing rocks at windows causes them to break. This is your scientific theory. But is it necessarily true? If there's any doubt, repeat the experiment. Repeat it a million times, if you like. But who's to say that on the million and first occasion the rock will still break the window?

The "scandal" of induction, as it's come to be known, is that observations are necessarily limited—you can't keep throwing rocks at windows forever. That is, the experimental method is "finitistic": It deduces truth from only a finite number of experiments or observations. But who's to say that, if carried out long enough, an experiment won't yield negative results? And by assumption if experiment even once contradicts a hypothesis, that hypothesis can't be *necessarily* true—that is, it can't be known with certainty.

This was more or less Hume's point about inductive knowledge: It's an oxymoron. The things we think we "know" from experience—that the sun will rise and that birds fly and so on—are really only things we *believe,* because we've become habituated to experiencing them. We believe that throwing rocks "causes" windows to break because we've observed this regular pattern; but it is impossible to observe causation itself. Cause is inferred or induced from the mutual occurrence of two events (rock is thrown, window breaks). Looking at the world this way—in terms of cause—however, is really just a mental habit. There's no way of proving or knowing for certain such a thing as "cause" or that anything that happens in the world must necessarily happen again.

Hume, however, saw no cause for despair in this scandal. If certain knowledge of the world is philosophically uncertain, that doesn't mean we can't know anything. It just means that we know things in a particular and limited way. At the very least, we can assert the great likelihood or probability that the sun will rise tomorrow and that a stone will break a window. And in fact, absent any good reason to doubt them, we may call such beliefs "certainties" in a practical (if not philosophical) sense. Furthermore, Hume believed that if our ideas lead us to believe in laws of nature, that is because nature really *is* lawful.

Nonetheless, Hume's successors, who thought knowledge must be certain to really be knowledge, weren't satisfied with "practical certainty." An apparent solution to the dilemma was proposed by Immanuel Kant, who rejected Hume's claim that matters of fact are known only by experience. To use his jargon, such concepts as causality, space, and time are "synthetic *a priori*" notions, which means that they describe reality but are not induced from experience.

The catch in Kant's solution is that it makes natural laws and other scientific truths dependent on mental categories, which is to say dependent on human consciousness. Strict induction never takes place, since only appearances, not things in themselves, are accessible to human knowledge. True, we may be certain about those appearances, and these couldn't exist unless supported by an underlying reality; but Kant still draws a line around ultimate reality and declares it unknowable.

Still not good enough. It would take a whole book to chart the manifold approaches to resolving the problem of induction, ranging from the utilitarian to the pragmatic to the phenomenological. But one philosopher in particular deserves mention, since his name has become synonymous with the twentieth-century critique of scientific method: the Austrian Karl Popper (born 1902).

Like Kant, Popper sought to eliminate the problem by eliminating induction altogether. In his view, scientific hypotheses do not arise from observation, but rather as a free creation of the imagination (individual or communal). We observe and experiment to *test* rather than create theories. And the rationale of testing (whether acknowledged or not) is, ironically, not to prove theories certain, but rather to prove them false. Testing a hypothesis would be meaningless if failure were not possible. And if contrary evidence is found, then the theory is refuted and another must be created. Theories that survive experiment are not, on the other hand, proved "true"; they have only been shown to give us a good working description of reality.

Furthermore—and this is Popper's most interesting point—the harder it is to falsify a theory, the less meaningful the theory. For obviously, the more general the statement, the harder it is to find exceptions. (For example, "Black swans exist somewhere in the world" is difficult to falsify, while "Black swans exist in Rhode Island" is easier.) The more content a theory has—the more precisely it attempts to account for phenomena—the more meaningful it is, but also paradoxically the easier it is to refute. But refutation is no calamity; it is, as we said before, the essence of scientific pursuit. (If a claim is not at least potentially falsifiable, then it isn't scientific, it's just obvious—or, to return to Hume's terms, it states merely a "relation of ideas.") Refutation leads to the refinement of old theories and the creation of new ones, so that our picture of the world becomes more detailed, durable, and complete. If we knew things certainly and irrefutably, then science would be finished.

The Thing-in-Itself (Das Ding-an-Sich)

From this it follows incontestably, that the pure concepts of the understanding never admit of a transcendental, but only of an empirical use, and that the principles of the pure understanding can only be referred, as general conditions of a possible experience, to objects of the senses, never to things in themselves....

Immanuel Kant, *Critique of Pure Reason* (1781)

The important thing to know about a "thing-in-itself" is that there's nothing to know. You don't even have a chance.

The reason, according to German genius Immanuel Kant (1724–1804), is that our minds never come into direct contact with ultimate reality. Because of the way our senses work, and because our brains are prefitted with various concepts and filters, the reality that we perceive and "understand" is at least a step or two removed from things in themselves.

Kant's "thing-in-itself" (*Ding-an-sich*) is buried deep in his daunting masterwork, *Critique of Pure Reason,* an attempt to correct the deficiencies in contemporary philosophies of knowledge. His main inspiration was the work of empiricist David Hume, which, according to Kant, awoke him from the "dogmatic slumber" of orthodox rationalism (the view that reason is our chief source of knowledge).

Kant came to believe, along with Hume, that knowledge stems originally from experience rather than reason. But he rejected the empiricist position that experiences directly imprint themselves on the brain, which is a pure "blank slate" at birth. According to Hume, no concept—not space, time, substance, causality, or any other mental category—is *a priori* (prior to experience).

Kant didn't buy this argument at all. In the first place, he said, Hume must be wrong about concepts of space and time. For Hume had claimed that we learn the concept "space" by observing relations among the objects we perceive. We see that this thing is next to that thing, or on it, or under it, etc., and come to understand that such relations take place in space. Likewise, the concept of time, said Hume, is derived from observing that events happen in a sequence (this thing happens after that thing and before this other thing, etc.). But Kant destroyed Hume's logic. How can we experience that one thing is "next to" another or that one even happens "after" another unless we *already* have concepts such as "next to" and "after"—that is, concepts of space and time?

Such things just have to be built into the mind, Kant concluded, or else we could never even begin to make sense of the chaos that is perception. Space, time, and a host of other "categories" (such as quantity, quality, relation, and cause) must be inherent to thought; they are the forms we impose on experiences to organize and understand them. And that everyone shares the same ideas about space, time, etc., means that in addition to being *a priori* (innate) they must also be universal.

On the other hand, they can't be said to exist in the same way objects do, for such intuitions and categories are only concepts, not things. We impose them *on,* rather than discover them *in,* experience. Experience, too, is different from reality, because it consists of sense impressions or perceptions of things, not of things-in-themselves. (It is because concepts apply only to experience that we cannot use them to describe transcendental objects— God, for example.) What we perceive in space and time, what we see as being this large and that color, what we deduce causes such-and-such an event, are only the *perceptible* aspects of things, the face they show to us, what Kant calls "phenomena."

If space and time exist only in the mind—as Kant certainly implies—then, by experiencing the world as existing in space and time, we are experiencing only how the world *appears* to us, not how it is. But certainly there is a real world out there supplying us with those appearances: Kant calls this the "noumenal" (as opposed to "phenomenal") world. This is the world in which a thing is just what it is, not what it appears; this is the world of the *thing-in-itself*.

By definition, we can never experience this world; how, then, do we know that it really exists? Kant was never able to do any better than suggest that we must believe in things-in-themselves if we are to believe in true knowledge, and this part of his theory was somewhat embarrassing to his successors. Every great philosopher after Kant attempted to do away with it, either by claiming that appearances *are* (or at least can be) the same as reality, or by claiming that, while we do live in a world of phenomena, philosophy can lead us to the noumena behind them. We shall meet up later with two philosophers who held the latter view: G.W. F. Hegel and Edmund Husserl.

The Categorical Imperative

There is ... only a single categorical imperative and it is this: Act only on that maxim through which you can at the same time will that it should become a universal law.

Immanuel Kant, *The Metaphysic of Morals* (1797), chapter 11

Having solved the problem of knowledge, at least to his own satisfaction, Immanuel Kant fed ethics into his philosophical machinery, and out came an update of the Golden Rule.

The original version simply states, "Do unto others as you would have them do unto you." Kant modifies this into something like "Do unto others as you would have everyone do unto everybody." In more technical terms, your actions should be based on principles you would wish were universal laws. This is what makes the law a "categorical imperative," to use Kant's term: categorical because it covers everyone without exception and imperative because it's a moral duty.

Why is this better than the original? Because in theory it avoids the problem of people's differing ideas about what they'd like done unto them. Kant hoped to counter what we now call "moral relativism," the notion that what is right or good depends on the situation or context. Particularly wrong, he thought, was the doctrine of utilitarianism—namely, that ends justify means. How can results provide a moral basis for action, he asked, when even the best-laid plans may go awry? The results of what we do are quite often not what we intended, so it is morally skewed to base our judgments on results. What we can reliably judge are the principles on which we act, the rules we use in making decisions.

And how are such principles to be judged? If we want to be objective, we must look for universals. None of this situational morality stuff; Kant is looking for solid and consistent rules on the order of "stealing is wrong." The rules must apply across the board.

Here's how the categorical imperative works: Given a choice, you must be conscious of the rule on which you act. Suppose a stranger comes up to you and insults your mother. Your first impulse may be to strangle him, but stop and examine your motives. If you're saying to yourself, "It is right to strangle a person who offends me"—this would be the "maxim" for doing so—you must then question whether such a rule is rational and whether you would *wish* it to be applied across the board. Considering that the result would likely be mass murder and chaos, you must reject it—it is *imperative* that you do so.

Kant's rationalistic update of the Golden Rule has its difficulties. First of all, normal human beings are rarely capable, in the heat of the moment, of stopping to consider the universal implications of their maxims. Second, rules are nice, but not always practical—even if we'd like to do the right thing, we aren't always able.

Third, it's not always easy to figure out what the appropriate "maxim" or rule of thumb is, except in the simplest of situations. Lots of different rules might apply, some contradicting the others. Everyone who feels torn when approached by a panhandler will understand what I mean.

"And the Life of Man, Solitary, Poor, Nasty, Brutish, and Short"

Whatsoever therefore is consequent to a time of War, where every man is Enemy to every man; the same is consequent to the time, wherein men live without other security, than what their own strength, and their own invention shall furnish them withall. In such condition, there is no place for Industry; because the fruit thereof is uncertain: and consequently no Culture of the Earth; no Navigation, nor use of the commodities that may be imported by Sea; no commodious Building; no Instruments of moving, and removing such things as require much force; no Knowledge of the face of the Earth; no account of Time; no Arts; no Letters; no Society; and which is worst of all, continual fear, and danger of violent death; And the life of man, solitary, poor, nasty, brutish, and short.

Thomas Hobbes, *Leviathan* (1651), Part I, chapter 13

You often hear these lines from Thomas Hobbes's masterpiece quoted as if he meant life *now* or life in general. Though *Leviathan* is kind of a downer, Hobbes isn't *that* grim. He meant that life without civil society would be "solitary, poor," etc., not that life is actually like that now.

True, the later Renaissance was no picnic, in England or elsewhere. People remember Shakespeare and Galileo, but forget that advances in knowledge and navigation also fostered skepticism and doubt. The comfortable science of Aristotle lay in ruins, as did the certainties of a unified Church. The God who ruled the world through kings and queens, their dominion guaranteed by "divine right," gave way to a more intimate and personal God, who spoke to the common man as well as to the king.

The English rebels who beheaded King Charles I in the mid-seventeenth century obviously didn't see him as God's mediator. While Hobbes (1588–1679) agreed that any government is the result of social consensus, he violently opposed the rebels' anti-monarchial philosophy and acts. Only a ruler with absolute power, he argued, can effectively restrain men from mutual quarrel and exploitation. (Thus the title *Leviathan,* a reference to the massive and overpowering creature found in the Bible.) To establish this premise, Hobbes conjures up a scenario of life without law.

Hobbes depicts man in his natural state as self-centered, self-interested, and insecure. He knows no law and has no concept of justice; he only follows the dictates of his passions and desires, tempered somewhat by the promptings of his native reason.

Where there is no government or law, men will naturally fall into contention. Since resources are limited, there will be competition, which leads to fear, envy, and quarrel. Men also naturally seek glory by beating up on others. Since in the broad view people are roughly equal in strength and intelligence, no one person or group can securely seize power. Thus conflict is perpetual, and "every man is Enemy to every man."

In this state of war nothing good can arise. While everyone concentrates on self-defense and conquest, productive labor is impossible. There is no leisure for the pursuit of knowledge, no motivation for building or exploration, no place for arts and letters, no grounds for society—just "continual fear, and danger of violent death." Thus the life of man in such a state is, in Hobbes's most famous phrase, "solitary, poor, nasty, brutish, and short."

Such a view, which is in keeping with the diffidence and despair of the age, conspicuously lacks any reference to God. In particular, it lacks any reference to God's role in government, which Hobbes saw as a human product. Governments arise when man, prompted by reason, seeks the long-term good of avoiding his desperate natural state of conflict and fear, in hopes of attaining peace and security. Men *choose* to recognize a common power so long as their neighbors do the same, because only such a one can maintain order. This power then has the obligation to maintain the common security; its agency is law, and its vehicle is incontestable force. For to the degree power is shared, conflict will arise.

This is why Hobbes thought monarchy the best form of government: only a Leviathan-like power that itself is beyond the law—and thus subject to no higher authority—can surely and effectively maintain the commonwealth. Of course, as Hobbes admits, kings will fall to quarreling with other kings, but he doesn't advocate a central world government. So long as things are stable at home, it's fine in his view for kings to seek glory abroad.

Unfortunately, Hobbes's ideas made virtually no one happy. He was too royalist for the social-contract crowd, and too taken by social contract for the royalists; what's more, his views struck many as atheistic, though Hobbes denied the charge. The fact is, his thought was too complicated and idiosyncratic for anyone to swallow whole, though it would influence philosophy and political science for generations. (Spinoza is one of Hobbes's notable disciples.) Hobbes himself grew weary of debate in his old age and contented himself with translating Homer.

The *Tabula Rasa*

> Let us then suppose the mind to be, as we say, white paper, void of all characters, without any ideas. How comes it to be furnished?... To this I answer, in one word, from experience.

> John Locke, *Essay concerning Human Understanding* (1690), Book II

Lest the Latin scare you, *tabula rasa* is actually one of the simpler great ideas. It just means "blank slate," which is the original state of the human mind, according to certain philosophers.

Among them was the English physician John Locke (1632–1704), who in his *Essay concerning Human Understanding* studied the origin of ideas and their relationship to reality. He proposed that all ideas come from experience and that knowledge is simply relations among ideas. This means we can't have any ideas until we have experiences, so in effect the newborn's mind is empty. Locke called it "white paper"; *tabula rasa* (literally "erased tablet") predates him and suggests, contrary to his doctrine, that something was once there to be erased.

The upshot of Locke's "white paper" is that not only are we born without concrete ideas, we also lack abstract concepts such as morality, God, and freedom. Such things must be learned, as language is, and they are learned either by experience or by reflection and reason. These views led Locke to reject idealism and the whole notion of innate ideas in favor of commonsense philosophy.

Though reason has its place in human understanding, Locke said, it doesn't dominate experience. Mind is *not* over matter, because matter, through experience, provides the mind with ideas. Our simplest and most basic concepts (such as "loudness," "hardness," and "sweetness") are furnished by sense, and all more concrete ideas are built upon them. Other ideas come to us through reflection, including awareness of our own thought processes;

"thought" itself as well as "perception," "belief," "consciousness," "doubt," and so on are furnished by reflective experience. That such ideas are simple, however, doesn't mean they're innate.

Locke's doctrine of *tabula rasa* derives mostly from simple logic. For example, if we were all born with an innate idea of God, then we'd all have the same idea of God. But of course we don't. Similarly, if we were born with the idea of moral right, we'd all agree on what is right and what is wrong. But we don't. Finally, analytic truths such as "whatever is, exists" and "$2 + 2 = 4$" are not ideas obvious to everyone—for example, children and idiots. Locke also thought the premises of rationalism—mind over matter—were much too complex to be useful or valid. Like Occam, he thought simpler is better, and any account of knowledge that doesn't require innate ideas is simpler.

Actually, while *tabula rasa* seems a simple idea, Locke's argument ends up rather complicated. In fact, he sometimes contradicts himself and is eventually forced to admit that certain "faculties" must be innate. Among these are the five senses and the capacity to reason, which do count as "ideas" in some circles. Whatever the difficulties of his argument, it did steer British philosophy into what remains its characteristic empiricism. He failed to convince the French, however, who by and large remain rationalists. Just another reason so many English worry over European union and the Channel Tunnel.

The Social Contract

Man is born free, and everywhere he is in chains. Many a one believes himself the master of others, and yet he is a greater slave than they. How has this change come about? I do not know. What can it render legitimate? I believe that I can settle this question....

[I]f men recover their freedom by virtue of the same right by which it was taken away, either they are justified in resuming it, or there was no justification for depriving them of it. But the social order is a sacred right which serves as a foundation for all others. This right, however, does not come from nature. It is therefore based on conventions. The question is to know what these conventions are.

Jean-Jacques Rousseau, *The Social Contract* (1762), Book I, chapter 1

"Man is born free, and everywhere he is in chains." Thus begins Jean-Jacques Rousseau's prognosis of the ills of his time. Modern society, characterized by selfishness, inequality, petty tyranny, and inauthenticity, betrays the natural state of man, which is to be free, open, and happy. (Obviously, Rousseau doesn't buy Hobbes's depiction of natural man as unhappy, violent, and selfish.)

Rousseau (1712–1778) wasn't calling for the overthrow of the "system." His target was mainly a belief, common to king, aristocracy, and many a man on the street: that their society and government were as they ought to be, because God or nature (or both) had made them that way. *Au contraire,* argued Rousseau; God created us natural equals, naturally good, capable of self-government, and essentially solitary. If society exists at all, it is because people, not God or nature, created it for mutual security and benefit. And what man has made, man may unmake.

Rousseau's ideal society, however, is not a free-for-all, because anarchy is even worse than oppression. What he promoted was a "social contract" (*contrat social*). Given that all men are natural equals, the ideal society is a consensual creation, an agreement or "contract" among all its members. And if such a society is to be authentic and good, it must preserve as far as possible the indi-

vidual's natural freedom, which is to say his right to self-determination.

But this is tricky to manage, since contracts by definition involve an exchange, in this case one of rights and freedoms. Rousseau's prescription, in fact, requires that each member of society "put in common his person and his whole power under the supreme direction of the general will" (chapter 6). Only thus will a society be formed that serves the common good rather than the desires of any one person or powerful group.

But how does this jibe with Rousseau's other requirement— that society not infringe on the individual's natural freedom? His answer is that there isn't a conflict at all. For true freedom is moral, and moral goodness consists in wishing for that which is good for everybody. Private interests, when they conflict with the common good, are simply immoral, so we are not naturally "free," by his definition, to pursue them. The "general will" established when we agree to his social contract is exactly the same as moral freedom, which everybody would acknowledge if they only knew what was really good for them.

Whatever you think of Rousseau's logic, it had truly radical implications—namely, that we have certain inalienable rights that no state or society can abridge. In the ideal social contract, the general will is sovereign and government rules only by the consent of the governed. The social contract grants executive power *conditionally,* and the people may dismiss a government whenever they so choose. It is such sentiments that earned Rousseau the compliment of being called by Edmund Burke the "insane Socrates" of the French Revolution.

Speaking of Socrates, the first references to social contracts of a sort are found, not in Rousseau, but in the dialogues of Plato. For example, in the *Republic* Glaucon refers to "a compact neither to commit nor to suffer injustice"; this is "the beginning of legislation and of covenants between men" (Book II). Ultimately, however, Plato dispatches with this theory, and with democracy to boot.

Closer to home, social contract theory was given its first serious treatment in the seventeenth century by Hobbes and John Locke. Hobbes, for instance, argued that reason and fear compel individuals to seek peace by joining in a "covenant," according to which they surrender various rights to a central authority. In return, they may go about their business secure in the knowledge that any evil done to them will be punished by the law. In Hobbes's version, men do not retain sovereignty; it is in fact imperative that they give it up to, say, a king, for otherwise they will be ever tempted to exert their power unjustly over others. But if such a king does things the people don't like, it's too bad, because the king is not bound by the contract, which only obtains among those who gave him authority.

Hobbes's general approach proved very influential—after Machiavelli, he was the second great figure in the development of political science—but his particular ideas were attacked from all sides and eventually ignored. More successful were the ideas of Locke, who in his *Second Treatise of Government* (1690) put forth the argument that while men need government, government is ultimately the servant, not the master, of society. The social collective is completely free to dictate and modify the terms of its governance, preferably by setting them down in writing. This is exactly what the Founding Fathers did when, in 1781, they composed and ratified the most famous of all social contracts: the United States Constitution.

The Dialectic

Dialectic can mean a number of things; for example, its original sense in Greek is simply "discourse." But it has meant mostly one thing since German philosopher Georg Wilhelm Friedrich Hegel (1770–1831) got his hands on it.

Hegel's dialectic is conveniently summed up in three words: thesis, antithesis, synthesis. (These weren't Hegel's terms, but no matter.) Every established concept or state of affairs ("thesis") eventually gives rise to a conflicting concept or force ("antithesis"), and when the dust settles something new and better than either (the "synthesis") will arise.

Thus human progress, which Hegel took as a given. It also explains continuity, for thesis and antithesis do not annihilate each other; rather, the best of each is preserved in the new synthesis. But this synthesis will have its own gaps and flaws, provoking a new antithesis, and so on and so forth. In Hegel's view, the march of progress is impossible without conflict.

To get a handle on what Hegel meant, let's take one of his best-known examples: the "master/slave relationship." Imagine a master and slave, or lord and servant, as our thesis and antithesis respectively. The master is a master, or the lord a lord, precisely because he keeps servants: He is defined, in this sense, by what he is not (a slave or servant). It might seem that in every way the master is superior to the slave, and that the slave owes his identity and well-being to the master. But all is not as it seems. For in fact, the master himself depends on the slave—not only for the services the slave provides, but even for his own identity ("I am master because my slave *regards* me as his master").

So the "thesis" master depends on the "antithesis" slave as much as the slave depends on the master—in a sense, the master is slave to the slave, and the slave the master's master. If the master reflects on this situation—as in Hegel's progressive view he is bound to do—he must come to realize the arbitrariness and injustice of subjecting the slave on whom he depends. And out of this realization will come a more just synthesis—a more rational and humane social arrangement.

History, philosophy, science, religion, and just about everything else work just like this, getting better and better all the time! But where did it all begin? According to Hegel, with a great immutable force he calls "Spirit" (*Geist*). This Spirit, which is ever pressing toward perfection, turns out to be Reason itself. Striking back at skeptics and other naysayers, he sweated out a

long and tedious argument to prove that Reason (=Mind= Spirit) is all-encompassing and ever striving toward total self-awareness, which is mankind's perfect fulfillment. And the medium through which Spirit strives is the dialectic, which is the unfolding of the "transcendental Idea" that will achieve its perfection at the End of History.

You might have guessed that behind all this mumbo jumbo is a much more familiar idea: God. This makes sense out of his contention that the phenomenal world is *created* by—is in fact just an extension of—*Geist*. What happened in the beginning, he muses, is that God—who is Absolute Mind—eventually turned His thoughts to Himself. But you have to set yourself apart from something to think about it—even when the object of thought is oneself, one must still be an "object" of one's own "subjectivity." Yet since God is All there is nothing apart from God. So what He thought was Nothingness.

Nothingness is of course the antithesis of Being. But rather than canceling Himself out, God set in motion a synthesis of Being and Nothingness, which is Becoming—the origin and central mechanism of History. And the aim of History is to lead back to Absolute Mind, to Being without Nothingness. And that's a *simple* version of his argument.

-Isms, -Ologies, and other Alarming Developments: Philosophy since the Nineteenth Century

Utilitarianism

"The greatest happiness of the greatest number," wrote Englishman Jeremy Bentham (1748–1832), "is the foundation of morals and legislation." Which is to say that actions ought to be geared toward producing happiness: "All's well that ends well" is the essence of utilitarianism. Or, from a critical angle, "The ends justify the means."

In fact, utilitarians such as Bentham and John Stuart Mill did promote a sort of moral relativism. Actions, in their view, can't be judged in isolation from circumstances and effects. A moral absolutist would say that murder is simply wrong, no matter the situation. A utilitarian, however, would say murder is right if it serves the greater good. Suppose you could have assassinated Hitler; a utilitarian would have told you to go right ahead, since it is better that *one* man die than many.

Since he equates good with what makes the most people the most happy, Bentham's version of utilitarianism is called "universal." (Not all utilitarians are universalists; some say that what's good for you is something only you can judge.) One problem with Bentham's approach, though, is that you can't always assure happy outcomes, and the best course of action might only be apparent in hindsight. You might have *meant* to please everyone by serving chocolate cake, but since everyone was either dieting or allergic you made them all unhappy. Do we judge your action as "wrong" despite your intention to bring the greatest happiness to the greatest number? According to utilitarians, yes.

A related difficulty is that not everyone agrees on what "good" means. Bentham's standard was happiness, but there are persuasive arguments to the contrary. For example, many would say that humility is a good, even though it can be painful. Likewise, ignorance is bliss, but it isn't everyone's idea of what is good or useful. Perhaps it's better to be unhappy now if that means you'll be happier in the future. But judgments like that depend on long-term results, and people might argue over how long the term should be. Is it good, for example, to get into a trade war with Japan now, and thus punish American consumers, if the long-term effect is a stronger American economy? Perhaps, but in the longer long term the benefits might disappear, while enmity between the countries remains.

Bentham made some interesting long-term plans of his own. He willed his library and an endowment to University College in London, where he'd taught philosophy. The catch was that Bentham's body had to continue attending faculty meetings. The skeleton was duly stuffed and preserved and in fact remains on display under glass at the college, though Bentham's head has been replaced with a wax model. (The original, which was beginning to decay, sits in a metal box at his feet.) Bentham very occasionally still attends meetings, where he is no doubt one of the more animated participants.

Superman *(Übermensch)*

And Zarathustra spoke thus to the people:

"*I teach you the overman.* Man is something that shall be overcome. What have you done to overcome him?

"All beings so far have created something beyond themselves; and do you want to be the ebb of this great flood and even go back to the beasts rather than overcome man?"

Friedrich Nietzsche, "Zarathustra's Prologue," *Thus Spoke Zarathustra* (1883)

Friedrich Nietzsche's *Übermensch* has undergone one of the most astonishing of high to low culture transformations—from stoic, unconventional man to Man of Steel. The comic-strip creation of Siegel and Shuster has little to do with Nietzsche's original aside from the name, and even that's a little off. *Übermensch* literally means "overman," not "superman," though the latter is used in the earliest English translations of *Thus Spoke Zarathustra,* and it was picked up by George Bernard Shaw for *Man and Superman.*

The character who preaches the "overman" is Zarathustra (a.k.a. Zoroaster), a Persian prophet of the sixth century B.C. who serves as Nietzsche's prophetic/poetic mask. Contrary to what you might think, Zarathustra isn't talking about a muscle-bound super intellect or any other "evolved" specimen of humanity. In fact, we all (*mensch* is gender neutral) are potential overmen. What it takes is not more workouts or smart drinks, but courage and will, for our biggest obstacles are fear and custom.

At the base of Nietzsche's concept is what he refers to as "the will to power," which he sees as the basic motivating force of all living things. "Power" doesn't mean brute force, or domination of others, but something more like "fearlessness." Since we're primarily motivated by the will to power, whatever we most admire or emulate must best represent power. Such (Nietzsche asserts) are self-harmony, self-control, and self-realization—exemplified, for example, by Socrates' calmly downing the cup of hemlock.

By and large, we fall way short of this *Übermensch* ideal. Far from being possessed of ourselves, we are motivated mostly by fear, habit, superstition, resentment, and everything else that makes up the "slave mentality." From birth, we are trained by family, church, and school to submit to rules and laws, act normal, believe superstitions, and enslave ourselves to various masters. Everything chalked up to "human nature" is really just custom. We grow both lazy and fearful of challenge and danger, numb to the stirrings of the inner conscience.

A world without customs or masters is frightening, but the alternative is slavery and alienation. We may give in to society, or we may conquer our fear and become *creators* rather than mere creatures. Creation is making one's own meaning of the world, which is in reality chaotic and ever changing; to be a creature is to submit to others' meanings. All language, in Nietzsche's view, is a more or less inspired interpretation, a kind of "lie." If language lies, one is better off—truer to oneself—making up one's own falsehoods. This is what it means to be "overman."

What it does not mean is to become just another slavemaster. Tyranny is the result not of self-mastery, but of frustration: It is simply the expression of more resentment. The overman, rather, lives life without resentment, and is in fact ready, if the situation requires, to serve as well as to lead. (For the rest of us, service just seems like slavery.) By controlling his passions and channeling his will, the overman creates art and philosophy, affirming life and laughter and all good things.

This would be a nice opportunity to get into another of Nietzsche's great ideas, which is the contest between the so-called "Apollonian" and "Dionysian" strains in Greek art, their relation to reason, passion, and will, their role in his theory of sublimation, and so forth. But that would take us pretty far afield, so here I leave you to consult my entries on those subjects in *By Jove: Brush Up Your Mythology!* (HarperCollins, 1992).

Eternal Recurrence

How, if some day or night a demon were to sneak after you into your loneliest loneliness and say to you, "This life as you now live it and have lived it, you will have to live once more and innumerable times more; and there will be nothing new in it, but every pain and every joy and every thought and sigh and everything immeasurably small or great in your life must return to you— all in the same succession and sequence…." Would you not throw yourself down and gnash your teeth and curse the demon who spoke thus? Or did you once experience a tremendous moment when you would have answered him, "You are a god, and never have I heard anything more godly."… [H]ow well disposed would you have become to yourself and to life to *crave nothing more fervently* than this ultimate eternal confirmation and seal?

Friedrich Nietzsche, *The Gay Science* (1882), section 341

Take a small collection of things, say the words in the sentence "Here is a small sentence." Feed them into a computer and instruct it to randomly rearrange the words—the results being something like, "is sentence here small a." Now suppose the computer were to repeat this process over and over. The output might look something like this:

1) here is a small sentence
2) is sentence here small a
3) small here a is sentence
4) a here is sentence small
5) sentence a small here is

…And so on and so on.

Now let's say the computer kept mixing up the words into infinity. Obviously, at some point it will start repeating itself, since there are only 120 possible permutations of the five words. What's more, given that it's got infinity to work on this little project, the computer must ultimately get around to repeating those same first five permutations in the same exact order. That is, sequence 1–5 will recur or "return," and in fact will do so over and over again, so long as the words stay fixed and the process goes on to eternity.

This word game is a boiled-down version of what Nietzsche was getting at when he came up with the doctrine of "eternal recurrence," a term that first appears in *The Gay Science,* a collection of allegories and aphorisms. He was as serious about this idea as he was about anything, considering it the most perfect of scientific hypotheses. Nietzsche presumed that the universe is composed of a finite number of "power quanta," whatever they are, arranged in finite space. Time, however, is infinite, and so, his logic went, these "quanta" must like the words in our sentence fall into patterns that repeat over and over in the course of eternity.

Nietzsche, in this regard a deterministic materialist, thought that all matter and every event could be expressed in terms of physical relationships, namely among his "power quanta." The world as we know it is just one particular arrangement of quanta, as are such facts as your date of birth and taste in music. The quanta keep combining and recombining to form new realities, which will eventually include the disintegration of the world. But never fear: Since time is infinite, the quanta must someday re-arrange themselves back into the world as we know it. In fact, given enough time, they will repeat *exactly* the arrangements that have produced human history. History, in short, will return, over and over again in the course of eternity.

Now for many people this very thought is hell, a depressing version of the movie *Groundhog Day.* But that's Nietzsche's real point: He proposes eternal recurrence to prompt us to consider what it *might* be like to have to live our lives over again. If you find the notion unendurable, then that says something—namely, that you're in the grip of the "slave mentality." But if you "crave nothing more fervently," then congratulations: You've become a "superman." (For instructions, *see* p. 56.)

Scientists, of course, take a pretty dim view of Nietzsche's theory. Spitting out word permutations in linear order is one thing, but arranging matter in three-dimensional, continuous space is another—even if you accept that space is finite and time infinite. It's quite possible, for instance, to set a small number of objects in motion so that they *never* repeat their initial position (or any other), even if let go on to doomsday and beyond. And this isn't even to question the bizarre concept of "power quanta," which doesn't stack up with the discoveries of modern physics or with Einstein's theory of relativity.

Though eternal recurrence might seem like bad science fiction to you, it actually has a very respectable philosophical lineage. The idea was common among the ancient Greeks; followers of Pythagoras, for instance, believed in a so-called "Great Year" after which the cosmic cycle would repeat itself more or less exactly. (Heraclitus measured the Great Year to be 10,800 Earth years.) Aristotle seemingly approved of the basic notion, which was also held by the Stoics. (Their version is in fact very close to Nietzsche's.) Judeo-Christian belief in one-way time, with Creation at the beginning and Judgment at the end, seriously crippled the doctrine, but didn't kill it. Medieval and Renaissance philosophers—even Descartes—grappled with the possibility of repeating cycles of history, by no means rejecting it out of hand. Nietzsche, though, made the theory fashionable once more, if only for a short while. If eternal recurrence holds good, I'm sure we'll be seeing it again.

Pragmatism

To attain perfect clearness in our thoughts of an object, then, we need only consider what effects of a conceivable practical kind the object may involve. … Our conception of these effects, then, is for us the whole of our conception of the object, so far as that conception has positive significance at all. This is the principle of Peirce, the principle of pragmatism.

William James, "Philosophical Concepts and Practical Results" (1898)

Philosophy does have its little ironies. Charles S. Peirce (1839-1914), credited by Henry's brother William James as the inventor of pragmatism, was anything but a practical success. Cashiered from Johns Hopkins and academically blackballed for alleged immorality, Professor Peirce was reduced to beggary, dying a poor and neglected man. (He had the last laugh on academe, however, by co-inventing semiotics—*see* p. 183.)

Peirce, a failure despite himself, was a polymathic genius who made real contributions to mathematics, philosophy, chemistry, psychology, and statistics. His most enduring legacy, though, is the philosophy championed by James and later John Dewey. Disdaining metaphysics—the study of intangible or abstract "realities"—Peirce insisted that the meaning of an idea lies solely in its practical effects. If an idea has no effect, it is meaningless—either "gibberish" or "downright absurd," to use his own words.

The upshot is that such concepts as "goodness" and "truth" have no reality prior to or apart from what we *do* with them—apart from how they affect our behavior. Belief, in the pragmatic view, is the same as action or at least potential action. If our idea of goodness includes helping old ladies across the street, then goodness, for us, amounts to a tendency to perform such acts. Goodness is, finally, the sum total of all such effects of the idea. James, in various works, extended this notion even to the concept of God. He wrote in *Pragmatism* (1907), for example, that "If the hypothesis of God *works* satisfactorily in the widest sense of the word, it is true" (my emphasis).

Since James's time pragmatism has assumed a multiplicity of shapes, not all of them amenable to the founders. Yet pragmatists do share a core of basic premises, such as flexibility in method, scorn of dogma, and a relativistic view of values. ("Good" and "bad" are defined according to particular human needs, desires, and practices.) Peirce himself, though, was not much of a relativist. He believed that the universe is governed by evolving laws, ever more regular and less chancy. He believed that science, too, is evolving, and that scientists are gradually and steadily approaching the "truth" of universal laws. Others would say that "truth" itself is just a relative concept defined by human needs and behavior. "The true," James wrote, "is the name of whatever proves itself to be good in the way of belief." The truth is just what works.

One of the more famous (or infamous) partisans of the latter view today is Stanley Fish, a professor of English and law at Duke University. Fish is fond of exposing the relativity of supposedly certain concepts, as in his most recent book (as of this writing), *There's No Such Thing as Free Speech, and It's a Good Thing, Too* (1994). Freedom of speech is, according to Fish, not something absolutely given either by natural right or by law; it is just what we communally agree to allow. Some things—like "fighting words" or pornography—will always be excluded, and the definitions of both "free" and "speech" are constantly modified to suit communal needs. Fish has in the past, along similar lines, also dissected literary theories, notions of "quality," and legal doctrines.

Fish is just one of a number of neopragmatists, the most influential of whom is erstwhile philosopher Richard Rorty. Rorty, realizing that pragmatism logically implies that studying philosophy is nonsensical, took the admirable step of changing careers.

(He's now an English professor.) Such contradictions, it should be noted, are by no means unique in pragmatic thought. Another good example is that of Steven Knapp and Walter Benn Michaels, protégés of Fish, who caused something of stir in literature departments by publishing a theoretical essay titled "Against Theory" (1982). The gist of this wonderful piece is that all literary theory is based on a distinction between what a text "really" means and what its author intended. But the distinction is unreal, since *in practice* the two are one and the same: Whenever we speak of "meaning," we can mean nothing else than "intention." Therefore, to quote the authors, "the whole enterprise of critical theory," being "incoherent" and practically ineffectual, "is misguided and should be abandoned." But nobody listened. The article had no practical effect. Theory continues to rule. So pragmatism should be abandoned.

"The World Is All That Is the Case"

Among the most revered of twentieth-century philosophers, Viennese aristocrat Ludwig Wittgenstein (1889–1951) has been claimed as a forefather by several philosophical schools. But not all agree on what he meant, which is ironic, since perhaps his most important point was that philosophy should concentrate on ordinary—and presumably clear and uncomplicated—language.

Wittgenstein more than anyone is responsible for the focus of recent philosophy on language, in particular on the relationship between statements and realities. He proclaimed his own view in the opening sentence of his first work, *Tractatus Logico-philosophicus* ("Treatise in Logical Philosophy," 1921): "The word is all that is the case." Which means, he explains, that the world "is the totality of facts, not of things."

If you don't get the distinction, it's this: "Facts" are true statements about things. A chair is a thing; the statement "The chair is red" is (or may be) a fact. The "world" as we know it is simply the collection of known facts—of "what is the case"—rather than of things separate from what we can say about them. It is language that constructs our sense of the world, our surroundings and experiences. What we cannot say we cannot know; "What we cannot speak about we must pass over in silence."

Wittgenstein's propositions deeply influenced a group of young philosophers known as "logical positivists"—believers, along with Hume, that whatever is neither self-evident nor empirically demonstrable is just nonsense. ("Nonsense" by their lights included literature, art, and fancy metaphysics.) But though they embraced Wittgenstein, Wittgenstein did not embrace them. For while he too thought philosophy ought to restrict itself to "what is the case," he remained haunted by silences and unknowable realities. What is not factual may be nonsense, but as far as Wittgenstein was concerned nonsense is very interesting.

Wittgenstein's views evolved between the *Tractatus* and the posthumous *Philosophical Investigations* (1953), which reconstructs lectures he gave at Cambridge. In fact, he all but abandoned many of his earlier views, such as the claim that the "limits of my language are the limits of my world." Like most philosophers of language, the young Wittgenstein had treated words as pointers to or representatives of things in the world. But the later Wittgenstein thought such an emphasis on reference too simplistic.

In *Philosophical Investigations,* Wittgenstein offers a new view: what words mean depends not on what they refer to, but on how they are used. Language, he said, is a kind of game—a set of "pieces" or "equipment" (words) which are used according to a set of rules (linguistic conventions). As in the *Tractatus,* our world is constructed out of statements or potential statements, but now the emphasis is less on what statements "mean" (denote) than on how, given rules and a context, they are deployed.

It follows from this that knowledge lies not in discovering (or inventing) some "reality" that corresponds to our speech, but rather in studying the way speech works. Thus ordinary language is the proper subject of philosophy. Traditional philosophical problems, concerning such concepts as "being" and "truth," are merely confusions that arise out of jargon and the misguided attempt to discover the "reality" it supposedly "represents."

Phenomenology

That we should set aside all previous habits of thought, see through and break down the mental barriers which these habits have set along the horizons of our thinking, and in full intellectual freedom proceed to lay hold on these genuine philosophical problems still awaiting completely fresh formulation which the liberated horizons on all sides disclose to us—these are hard demands. Yet nothing less is required. What makes the appropriation of the essential nature of phenomenology... and its relation to all other sciences... so extraordinarily difficult, is that in addition to all other adjustments *a new way of looking at things* is necessary, one that contrasts *at every point* with the natural attitude of experience and thought.

Edmund Husserl, *Ideas* (1913), Introduction

Lots of philosophy is pretty unreadable, but the writings of Edmund Husserl (1859–1938) deserve some sort of prize. Jargon-laden and obsessive, Husserl's phenomenology has nonetheless left a major mark on twentieth-century philosophy. In fact, the great indigestible works of Heidegger, Sartre, and Derrida wouldn't have been possible without him.

The terms *phenomenon* and *phenomenology,* from the Greek for "appearance," were not new to Husserl. Already common in German philosophy, *phenomenon* refers to a thing or event as it appears to human consciousness (as opposed to what it is essentially, apart from perception). *Phenomenology* is thus the study of manifestations. Husserl believed that as far as our knowledge of the world goes, all we have are phenomena so we should try to make the best of it.

To do this, we must begin by stripping our perceptions down to their simplest forms, shedding all our layers of habit and assumption. We can train ourselves to look at, say, a chair without any thought of its purpose (for sitting), history (who built it and where we bought it), or function (whether it's comfortable or matches the rug). We must try to just experience the chair in the most immediate way, as a pure object of consciousness. When we see it in this way—grasping only its most essential features—the chair has become a "phenomenon" in Husserl's sense.

Husserl called this kind of perception "bracketing," or, in Greek jargon, *epochē,* which he defined as "suspending belief in the existence of objects." He meant, roughly, that in searching for the essential features of a thing we must suspend any attachment to its actual existence. We then perform something called "free imaginative variation." Take our chair; imagine how it might be different—mentally chop off a few legs, reverse the back, add a fax modem. Is it still a chair? Whenever we feel that we would not intuitively recognize the resulting variation as a "chair," we know we've changed something essential. This sort of "bracketing" is the means to get at the essence of a thing apart from how it exists here and now. "Back to the things themselves" was Husserl's motto, and he meant "back to our primal intuitions of things in their essential form (as phenomena)."

But phenomenology doesn't stop there, because we've yet to arrive at the most basic phenomenon, which is consciousness itself—the faculty of having intuitions. No matter how good we get at bracketing chairs, we still don't understand the structure of even simple conscious experience. To examine consciousness, we need to somehow get beyond or above it and achieve what Husserl called "transcendental subjectivity." (Normal people call this "self-consciousness," but while we think of that as subjective, Husserl's "transcendental subjectivity" is supposed to give us an "objective" view of consciousness.)

I'll leave aside Husserl's technical explanation of exactly how we escape subjectivity. Suffice it to say that if we examine consciousness, the first thing we notice is that it is "intentional." In Husserl's sense, "intentional" means something like "directed"; we are never simply conscious, but always conscious *of something.* (This "something" can be an object like our chair, or a feeling such as hunger, or an idea, and it doesn't have to be "real.") It follows that since consciousness is necessarily directed, it is never empty or passive but always active. Consciousness reaches out and grabs things, and the way things appear to us depends both on their own character (their essence) and on the character of the "grabbing." We have a hand in making the objects of our world—for example, by giving them meaning.

The ultimate point of all this is to be able to describe, as fully and correctly as possible, the most basic ways in which we make sense and meaning of the world we live in. A noble aim, certainly, but a controversial one. Even some phenomenologists have had trouble with the notion of a "transcendental ego," especially as in his later years Husserl started making outlandish claims about it. (For example, he came to view the transcendental ego as an entity separate from one's consciousness and somehow immortal.) Is there really any point from which we can regard and describe the normal workings of the mind? How can consciousness transcend itself? I'll leave the solutions to these puzzles to you.

Existentialism

Existentialism? Why, Jean-Paul Sartre, of course. Except that the idea is a century older than Sartre, its best-known proponent.

Where existentialism—a study of subjective experience—really begins is with Danish philosopher Søren Kierkegaard (1813–1855). Kierkegaard accused contemporary philosophy of wasting too much time on "essences," the supposed underlying realities and universal laws of the world. Not only are such things dubious, focusing on them diverts attention from real problems, such as how we as individuals can make decisions.

In the first place, Kierkegaard rejected the idealistic belief that good and bad have some objective or essential reality. They are rather "subjective truths" which, though they can't be proved or extended to others, are the sole basis of individual actions. For example, we cannot say that killing is "bad" in some objective or

logical way; there are situations, indeed, when it is considered "good" (in self-defense, say, or in war). Most of the time, there is no way to logically decide on the right course of action. You cannot calculate how to respond to injustice, or whether to believe in God; but neither can you avoid decisions or beliefs.

There are of course some objective truths: two plus two does equal four, and Napoleon was defeated at Waterloo. But so what? As far as Kierkegaard is concerned, such truths—interesting as they may be—have no bearing on one's day-to-day existence or on one's crucial decisions and acts. What we *are*, he believed, is what we *do*. If we are truly to be, we must act, and we base our actions on our values—purely subjective and individual truths, articles of faith, unprovable but supremely real.

Neither nature nor society can offer us certainty about good and bad, right and wrong. The ultimate meaning and value of our actions are always uncertain. What it means to be human is to act in the face of such uncertainty. Men and women, in the existentialist view, act inauthentically if they behave merely as society bids them, or just accept the dictates of the church or any other institution. To do so is to evade responsibility.

The problem with the existential view, which is compelling so far as it goes, is that emphasis on individual existence and choice does nothing to address widespread social dilemmas. For more on this problem and one response to it *see* the following entry.

"I Am Condemned to Be Free"

Indeed by the sole fact that I am conscious of the causes which inspire my action, these causes are already transcendent objects for my consciousness; they are outside. In vain shall I seek to catch hold of them; I escape them by my very existence. I am condemned to exist forever beyond my essence, beyond the causes and motives of my act. I am condemned to be free. This means that no limits to my freedom can be found except freedom itself or, if you prefer, that we are not free to cease being free.

Jean-Paul Sartre, *Being and Nothingness* (1943), Part Four

When someone reminds you that "It's a free country," you know what he means. You're generally free to do what you want (this is called "positive" freedom), and generally free from persecution for your views ("negative" freedom). Positive freedom involves choices; negative freedom, consequences.

These twin freedoms are quite wonderful and we're lucky to have them. But the important word here is *lucky*. If in the unlikely event that a dictator were to seize control tomorrow, our prized freedoms could be abolished in a second. What would be left? Is there an essential kind of freedom that can never be taken from us?

According to Jean-Paul Sartre, this century's leading existentialist philosopher, the answer is yes. But that "yes" is a mixed blessing. Sartre says that to be human is to be absolutely free, to always have the power of choosing. But the one thing we can't choose is to renounce choice, or, to quote Sartre's paradox, "I am condemned to be free." Choosing not to act is still a choice. This is the existential dilemma.

Sartre's philosophy of freedom derives from his study of phenomenology, the philosophy of pure consciousness. As he saw it, what distinguishes consciousness is that it is at once of the world and not of the world. When we reflect on how we think, when we become self-conscious, we treat our thought as if it were an object in the world. To say "I was confused by that explanation" is to transcend our own thinking and reflect on it. But the world as we know it is just the collection of all such "transcendent" objects: things we perceive and think about.

At the same time, consciousness is not of the world. When we dream, we are cut off from any outside sensorium. When we imagine—say, when we fantasize about winning the lottery—we rise out of the present (the world as it is) and project a better future (the world as it is not). This future, because it isn't actual, is nonexistent: it is *nothingness*.

According to Sartre, all action arises out of this nothingness. If you were always directly attuned to the present, unable to escape it, you not only couldn't imagine, you couldn't act. The present is just what it is, and unless you consider how things might be different, there's no motive to do anything. Sartre took this thesis a step further: All our actions are directed at a goal that does not exist in the here-and-now. Our actions, then, being based on nothing, are never *necessary* either. Goals are things we freely create for ourselves, and along with them we create our own values. (Here Sartre is adapting Kierkegaard.)

Sartre's famous "nausea" arises out of absolute freedom of choice, the awareness that you are capable of any possible action whatsoever. For example, at any moment you may choose to kill yourself; and this very thought—which opens up a yawning abyss in the self—generates angst and nausea. (You *could* do it, and you dread that you therefore *might* do it.) To be "condemned to be free" is to be solely responsible for making out of each situation our own "world"—for choosing our own goals, our methods of coping, our responses to the anxiety of choosing. Perhaps you'll choose to kill yourself; otherwise you at least choose to keep on choosing.

Most people refuse to face up to these facts, however, because they can't stand the idea that they are *responsible* for their world.

As many a critic has said of our time, we would rather see ourselves as victims than as accountable adults. We blame our poor choices or failed endeavors on an unhappy childhood, on cultural oppression, on class, on prejudice, or on society in general. Sartre wouldn't deny that unhappy childhoods and cultural prejudices exist and are bad. But he labels as "bad faith" the refusal to own up to our free choices in interpreting and responding to the facts of life.

Sartre was very good at exposing bad faith, but less good at paving the way to authenticity; existentialism is better at describing than prescribing. Sartre himself eventually recognized the limitations of existentialism and grew more concerned with oppressive situations. By the late 1950s he had become enchanted with Marxism, not as a political system so much as a philosophy of collective action. Of course, the Marxist ideal of coordinated world-making hasn't turned out so well either.

SCIENCE AND MATHEMATICS

What Goes Up Must Come Down: Basic Theories

The Pythagorean Theorem

Thales thought the world is made of water; Heraclitus said the primal stuff is fire. Pythagoras (sixth century B.C.) thought all is number, and equations the path to all truths. "Things are numbers" was his motto. Pythagoras's philosophy, which included beliefs in reincarnation and the evil effects of beans, is at times bizarre and obscure, but he did concoct the first great geometric proof, which is of the formula still known as the "Pythagorean theorem."

The theorem will be familiar to you if you've ever studied geometry. It concerns that most interesting of all triangles, the right triangle (one that contains a "right" or 90° angle). According to Pythagoras, all right triangles have a common property—namely, that the length of the longest side, when squared, is equal to the sum of the squares of the other two sides. In this picture, the length of the longest side (called the "hypotenuse") is c. According to the Pythagorean theorem, $a^2 + b^2 = c^2$.

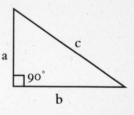

Actually, though ancient testimony unanimously traces the proof of this theorem to Pythagoras (a Greek from the island of Samos), the theorem itself was kicking around the Middle East for about a millennium before his birth. And however he proved it, his proof is not extant; the first record of it is found in Euclid's famous *Elements* (ca. 300 B.C.), which to this day forms the basis for elementary geometry. Euclid's version, in brief, requires constructing squares on each of the triangle's sides and demonstrating that the areas of the smaller two add up to the area of the largest.

Besides being the earliest of its kind, Pythagoras's proof is important for other reasons. First, it contributes greatly to the suffering of schoolchildren everywhere. Second, it eventually led to the discovery of "irrational" numbers—that is, numbers which cannot be expressed as a fraction (as the *ratio* of two whole numbers). An example of an irrational number is the square root of 2, which is the length of the hypotenuse when the other sides of a right triangle are one unit long.

Ironically, this consequence of the theorem did not fit the Pythagorean program, since Pythagoras had believed that all numbers are rational, or, to use his term, "commensurable." (According to legend, other members of the Pythagorean school took the man who deduced incommensurable numbers and threw him in the Mediterranean.) As Greek arithmetic dealt only in rational numbers, geometry was thus seen as more powerful and better able to map the world. This is why Greek mathematics would concentrate on, and make great strides in, geometry while making very little progress in arithmetic.

"Eureka!" (Archimedes' Principle)

According to the story, Archimedes, as he was washing, thought of a way to compute the proportion of gold in King Heiron's crown by observing how much water flowed over the bathing-stool. He leapt up as one possessed, crying *heurēka!* ("I've found it"). After repeating this several times, he went his way.

Plutarch, "The Impossibility of Pleasure according to Epicurus"

The Sicilian mathematician Archimedes (ca. 287–212 B.C.) was the classic absentminded professor, a brilliant thinker often oblivious to the real world, especially after making one of his great discoveries. By Plutarch's account, when Archimedes discovered the principles of hydrostatics (the science of how solid bodies behave in liquids), he gave a couple of shouts and then ran off naked through the streets of Syracuse without explaining the fuss.

The fuss began when Archimedes' friend King Heiron II of Syracuse grew suspicious that a new crown he'd ordered was not solid gold, but gold alloyed with silver (or something worse). Short of melting the thing down, Archimedes knew of no quick way to determine the crown's composition. But one day, as he stepped into his tub, which was full to the brim, Archimedes had a brainstorm. "I've found it! I've found it!" he crowed—*heurēka!* (or *eureka!*) in Greek.

If Archimedes didn't coin the phrase—it's just a common Greek predicate—he made it famous. And he did discover the principle named after him, namely that one can determine the density of an object "O" by comparing its weight to that of the water it displaces in a bathtub. (The weight of the water, which has the same volume as O, is called O's "buoyancy"; the ratio of O's weight to that of the displaced water is called O's "specific gravity.")

Combining this discovery with the fact that mass equals volume times density, here's how to solve Heiron's riddle: Take a lump of pure gold weighing exactly as much as the crown in dispute. Drop the lump in a tub of water and measure (either by weight or by volume) the amount of water it displaces. (This would be the amount of water that spills over if the tub is full, or the amount the water rises in the tub if it isn't full.) Repeat the process with the crown. If both lump and crown displace the same amount of water, then both have the same volume, and the crown is pure gold, since no other metal is exactly as dense as gold. If, however, the crown displaces more water, it must be composed of gold alloyed with a less dense metal—its volume would be greater than that of the lump.

As it turns out, Heiron had been swindled: The crown was more voluminous than the lump. The jeweler was never heard from again, but Archimedes went on to have several more great ideas. For instance, he approximated the value of π, figured out how to compute the area of a circle, laid the foundations of calculus, and derived the theory of the lever. Especially pleased with the last of these discoveries, Archimedes boasted, "Give me a place to stand, and I will move the earth."

He never did manage this trick, nor did he successfully complete his calculation of how many grains of sand it would take to fill the universe. It is reported that he died while tracing a geometric diagram in the dust, as Rome was conquering Syracuse. According to Plutarch, so absorbed was Archimedes in speculation that he didn't hear the command of a Roman soldier to rise; the soldier, infuriated, ran him through.

The Copernican Revolution

It is only a myth that Columbus, by crossing the Atlantic, proved to a shocked world that the earth is round. In fact, few educated people since ancient times were flat-earthers; more troublesome was the question of whether this round world moves.

Astronomers from Plato's time through the sixteenth century tended to think that the earth sits still while the heavens rotate around it, but by no means were alternative theories lacking. In the fourth century B.C., for example, the Greek astronomer Aristarchus of Samos set forth what is known as the "heliocentric" hypothesis: that the planets, including Earth, revolve around the sun.

The trouble was that advocates of heliocentrism lacked any terribly convincing evidence. And such advocates were in any case few, especially after the second century A.D., when the great Greco-Egyptian astronomer Ptolemy devised a complex set of geometric equations to support the "geocentric" idea that the earth is the fixed center of the universe. Ptolemy's model and its formulas gave a seeming scientific basis to what people, by virtue of tradition, religion, and psychology wished to believe anyway. We don't experience any motion of the earth, after all, and the heavens do appear to circle the world; it is a pretty scary thought that our senses could be so deceived, and the apparent order of things be false. Christians, furthermore, found geocentricism conformable to their belief that the earth, and especially man, was God's most important, most central creation.

Finally, the weight of ancient science was on Ptolemy's side. Aristotle, the last word in science for nearly two millennia, proposed that the heavens are made of entirely different stuff from earthly things. This stuff, called *aether* or *quintessentia,* was perfect and incorruptible, and furthermore naturally moved in circles. Natural objects here on Earth, contrarily, were composed of the elements earth, air, fire, and water, which tended to either rise or fall. The elements earth and water fall, and Earth was made up of all the earth and water that had fallen to its proper place and was thus content to sit still.

Geocentricism, in short, was not simply a scientific hypothesis, but part and parcel of a complex, comforting, traditional world-view. But it is a view that finally met its match in the Renaissance, beginning with the work of the Polish astronomer Mikolaj Kopernik (1473–1543), better known by his Latinized name Nicholas Copernicus. Sent by a wealthy uncle to the best Italian universities, Copernicus was more a man of the book than a star-gazer. Ironically, he was particularly devoted to Ptolemy, but not so devoted that he could ignore the many failures of Ptolemy's geometry to explain the actual behavior of heavenly bodies.

A true Renaissance man, Copernicus was not quite the experimental scientist. As presented in *De revolutionibus orbium coelestium* (*On the Revolutions of the Heavenly Spheres*, 1543), his heliocentric theory—the first to gain any real attention in modern times—was still motivated more by metaphysics than by solid data. He did wish to find a model of the universe that would allow for more accurate predictions than Ptolemy's, but he also sought spiritual explanations, such as that the sun, as light-giver, is closer to perfection and to God than is the earth. (This argument would recur in the work of Copernicus's great successor, the German Johannes Kepler, 1571–1630.) And Aristotle lingered in his theory; he assumed that planetary orbits are perfectly circular, which they aren't. So Copernicus's model didn't yield results much better than Ptolemy's.

Nonetheless, the revolution had begun. Geocentricists such as Danish astronomer Tycho Brahe (1546–1601) fought back by refining the Ptolemaic model, while heliocentrists such as Brahe's assistant, Kepler, sought to simplify and adjust the Copernican system. The stakes were great, for if Copernicus was right and the earth was just one planet among others, then Aristotle's theory of *aether*, and along with it most of Aristotelian science, was wrong. And if Copernicus was right, the universe was much, much larger than had been supposed, for earlier calculations were based on Earth's circumference, while new ones would be based on Earth's orbit. Man and his world, it would turn out, are even more insignificant in the grand scheme than medieval scientists and theologians had thought.

When it came to explaining observed phenomena—or, in the parlance of the day, "saving the appearances"—neither Copernicus nor Kepler could do better than offer theories that were as good as Brahe's. In terms of pure math, it was a toss-up. That the heavens were made of the same stuff as Earth, and thus operated according to the same natural laws, awaited proof in the work of the great English physicist Isaac Newton. It turns out that Kepler had discovered several laws of heavenly motion that fit perfectly into Newton's new law of gravity [see NEWTON'S LAWS, p. 84]. Though this didn't prove for certain either heliocentric astronomy or Newtonian mechanics, it was a powerful argument for both, and a corner had been decisively turned. The heavens had been brought down to earth.

"Knowledge Itself Is Power"

This little aphorism appears in *Meditationes Sacrae* (1597), an obscure work by Francis Bacon (1561–1626), lawyer, politician, essayist, and co-inventor of the scientific method. On its face, the saying is obvious, especially in this age of information. But we're apt to misunderstand what Bacon means by "power," which is not "personal or political advantage," but "control of nature."

Bacon was campaigning against the sterile science and philosophy of his day. Scientific debate, chained to Aristotelian metaphysics and plagued by hair-splitting and sophistry, produced little save grounds for further debate. Meanwhile, the mechanical arts, which the theoreticians considered ignoble, had been making steady and swift advances; gunpowder, Gutenberg's printing press, and the mariner's compass were unmatched by progress in loftier realms.

Sizing up the situation, Bacon concluded that knowledge could be fruitful only if technology and philosophy were united. Instead of debating the fine points of matter and form, scientists ought to directly observe nature, draw conclusions, and employ practical tools to test them. In other words, science ought to be based on induction and experiment, not metaphysics and speculation.

Bacon was hardly the first to suggest the experimental or "scientific" method. And despite all his talk about it, he performed very few significant experiments of his own. Nonetheless, his contemporaries were impressed, and most great scientific minds of the seventeenth century, including Newton, cited his work as a direct inspiration. Furthermore, the collaborative character of scientific research from the 1600s to the present owes much to his repeated insistence that communities, rather than isolated geniuses, are responsible for true scientific progress and thus "power" over nature.

On the other hand, beyond his own practical shortcomings Bacon's theories do leave something to be desired. He tossed the baby out with the bathwater of speculative science, slighting the role of hypothesis, which he viewed as groundless and thus sterile. All true knowledge, he asserted, derives from observation and experiment, and any sort of prior assumption is only likely to distort perception and interpretation. But without hypotheses, there can be no controlled experiments, which are the essence of the modern scientific method. Bacon thought the world was essentially chaotic, and that therefore it was a mistake to approach nature assuming regular laws. But science has advanced principally by assuming the world is lawful, that there are regular and simple patterns underlying nature.

So Bacon got some things right and some things wrong, and on the whole he was much better at criticizing the old than at forecasting the new. As a result, his reputation has had its ups and downs. Current opinion is mixed; some celebrate his pioneering work in scientific philosophy, while others blame his doctrine of "knowledge is power" for skewing science toward the exploitation of nature. Power, in the view of these critics, has become an end in itself, resulting in materialism and alienation. Bacon himself thought that social values and morality would always direct and constrain technological advances. It is in this regard that he was most wrong.

Newton's Laws

Law I: Every body continues in its state of rest, or of uniform motion in a right [i.e., straight] line, unless it is compelled to change that state by a force impressed upon it.

Law II: The change in motion is proportional to the motive force impressed; and is made in the direction of the right line in which that force is impressed.

Law III: To every action there is always opposed an equal reaction; or, the mutual actions of two bodies are always equal, and directed to contrary parts.

Sir Isaac Newton, *Philosophiae Naturalis Principia Mathematica* (1686)

English physicist Sir Isaac Newton (1642–1727) is often called the greatest scientist in history; no doubt he's right up there. It was he who finally freed physics from metaphysics by formulating the universal laws of force and motion that guide both the heavens and the earth. But others had paved the way; as Newton himself said, "If I have seen farther than others, it is by standing on the shoulders of giants." (He wasn't the first to say that, either.)

One of Newton's predecessors was Galileo, the first great experimental physicist in history and the man most responsible for overthrowing Aristotle as the intellectual overlord of science. Galileo showed, for instance, that Aristotle's theories of motion were wrong, and that contrary to the widely held belief, objects do not fall at constant speed, nor do heavier bodies fall faster than lighter ones, at least in a vacuum.

That Galileo reached these conclusions by dropping objects from the Tower of Pisa is unfortunately a myth, but he did perform some clever experiments with inclined planes and the like. The result was his theory of uniform acceleration: falling bodies all gain speed at exactly the same rate. Galileo also discovered the property of inertia: bodies tend to either remain at rest or in constant motion unless disturbed by some outside force. From these two laws he was able to show that under ideal conditions projectiles follow parabolic paths.

A generation later, René Descartes—best known for his proclamation, "I think, therefore I am"—came along with the first "conservation law" of physics. Such a law states that some quantity remains the same despite a physical event or change of situation. Descartes was particularly interested in collisions of moving bodies—for example, of two billiard balls—and he proposed that their combined *momentum* (weight times speed) remains constant. That is, if you add up the momenta of two billiard balls after a collision, it would be the same as their combined momenta before the collision. Momentum is "conserved."

The discoveries of Galileo and Descartes were important, even revolutionary, but it would take Newton to assemble them into an interlocking system. The key concept joining Galileo's law of acceleration and Descartes's work on colliding bodies was gravitation, whose discovery is Newton's best-known achievement. (But you can put the apple-falling-from-a-tree-on-Newton's-head legend in the myth file along with the Tower of Pisa.)

Newton did not start right off with gravitation, but rather with another new idea, that of *mass* (*m*). Descartes had spoken of momentum as the product of weight and velocity, but Newton realized that weight is an imprecise and variable quantity—things weigh less in water, for example, than they do in air. He preferred something that could be fixed more exactly, the "quantity of matter" in an object—which would be the same everywhere, in water or in air, in space or on the ground—and this he called "mass." By substituting mass for Descartes's weight, and by treating velocity as a vector quantity (as speed in a certain *direction*), Newton arrived at a new definition for momentum (mass times velocity in a positive or negative direction). Since velocity is directed, even change in direction involves a change in momentum.

Newton then defined the quantity "force"—an old but vague concept—as that which causes change in the momentum of an object. Taking Galileo's law of inertia as his own first law of motion, Newton then postulated a second: that force (F) is directly proportional to the change in momentum it induces. In other words, twice as much force will cause twice as much change in the momentum of an object.

Given Newton's new definition of *momentum,* which substitutes the fixed quantity mass for the variable quantity weight, his second law amounts to the assertion that force is proportional to change in velocity. (If mass is fixed, extra force will affect only velocity.) Since change in velocity is the same as acceleration (a), F is proportional to a. Different objects, though, will require different forces to counter their inertia: The more massive an object, the more force is required to change its momentum; so F is also proportional to m. Thus we arrive at the famous formula embodied in Newton's second law: $F = ma$ (force equals mass times acceleration).

The most important consequence of the second law is that it allowed for the idea of a *continuous* force. Before Newton, scientists did have a concept of force, but only as something that could be communicated instantaneously, as in the example of the colliding billiard balls. Newton's force covered a much wider range of situations. A man pushing a wheelbarrow down the road, for example, exerts a continuous force to keep it in motion (that is, to counteract friction). The second law also opens the way for the idea of potential force, that is, the energy latent in a body held in suspension: The New Year's ball suspended above Times Square, for example, has potential force—an implied *future* rate of change (activated at midnight).

The last piece of the puzzle is Newton's third law, which is also his boldest leap. Taking off from Descartes's law that momentum must be conserved, Newton showed that whenever an object's motion is disturbed (whenever its momentum is changed), the motion of another object must also be disturbed so that momentum will be conserved. In fact, this second disturbance must be exactly equal to the first, but in an opposite direction.

Look at it this way: If we increase the momentum of a billiard ball by four units (what sort of units isn't important here), Descartes's law requires that something else must lose four units of momentum (or, put another way, it must gain *negative four* units of momentum—either by decelerating or moving in an opposite direction). This is the famous "equal and opposite reaction" Newton's third law requires. *Action* means change in motion, and thus change in momentum; and what the third law says is that to change the motion of one object, the motion of another object must also be affected—another object must intervene (it must suffer an equal and opposite change).

Here's where things get really interesting. Building on Galileo's law of uniform acceleration, Newton deduced the existence of gravity. As an object accelerates in its fall to the ground, its momentum increases. By the law of inertia and Newton's second and third laws, some force must be responsible for the acceleration, and it must be constant if acceleration is constant. ($F = ma$; the mass of an object remains the same, so if acceleration is constant, force must be too.) This force is exactly gravity, and Newton formulated the law that the force of gravity upon any object is constant and directly proportional to the mass of the object. (Since acceleration due to gravity is the same for all objects, F must increase in direct proportion to m.) The force of gravity upon a given object is equal to its *weight*. (Technically, weight equals mass times gravity; and, in case you're interested, the force of gravity is approximately 9.8 meters per square second.)

Even more interesting are the implications of Newton's third law. As an object falls to the ground by virtue of gravity, its change in momentum must be offset by an equal change in some other object's momentum. Let's say an apple falls to earth. The only other object involved whose momentum might be affected is the earth itself. Which is to say that the earth's force upon the apple must be equaled by the apple's force upon the earth; it is only because the earth is so much more massive than the apple that we don't perceive any change in Earth's momentum.

But in the world of abstract physical laws, which object is which is a matter of indifference. When an apple falls to the ground, we say that Earth's gravity makes it do so and ignore what happens to the earth in the process, because we don't perceive the equal and opposite reaction. But from the point of view of physics, it's just as good to say that the apple's gravity is pulling on the earth, and that the apple's fall is the equal and opposite reaction. From this fact Newton realized that gravity must be proportional to the mass of both objects involved—apple and Earth.

The practical effects of this revelation, which are negligible here on the ground, make a much bigger difference when we're talking about heavenly bodies, such as the earth and the moon or the sun and the earth. In fact, Newton's law of gravity, in an expanded form, finally proved that the solar system is indeed a solar system—that is, that the earth revolves around the sun rather than vice versa. Taking the data and laws postulated by the astronomer Johannes Kepler, Newton related the behavior of the planets to the equations of gravity, showing among other things that observed planetary motions are explicable only if every planet, guided by the sun's gravity, observes a slightly elliptical orbit around the sun.

Proving this now-obvious fact was just one of Newton's amazing achievements, which earned his theories, along with their assumptions, the status of absolute physical truths. But over the years one aspect of the theories would prove disturbing to many—it had even disturbed Newton. This was the fact that gravity apparently acts at a distance and with equal force. Until

Newton's day it was assumed that forces must be communicated by direct contact—as when two balls collide. But if gravity acts across distance, even in a vacuum, it must have nothing to do with any matter in between the objects involved. This just didn't seem sensible, and in fact it really isn't.

Only many generations later would a better explanation arise —the theory of fields, which we unfortunately can't get into here. Field theory in turn made Einstein's discoveries possible [*see* RELATIVITY, p. 92], and since then the Newtonian universe has crumbled. Newton's laws are still good when it comes to certain sorts of physical interactions, including most observable ones. But Newton's mechanistic universe is now known to have chance and uncertainty at its core, while his picture of gravity has been replaced by that of curved space-time. Which is a pity in a way, because Newton's world is so much more comprehensible than Einstein's and Heisenberg's.

Paradigm Shifts

Scientists like to think that they contribute to the steady march of progress. Each new discovery corrects deficiencies, making knowledge ever more perfect and truth increasingly clear. They look back on the history of science and discern a continuous development, conveniently marked off by the great discoveries.

But this picture is an illusion, according to historian of science Thomas Kuhn in his book *The Structure of Scientific Revolutions* (1962). Science is not a smooth transition from error to truth, but a series of crises or revolutions, expressed as "paradigm shifts."

By "paradigm" Kuhn means a set of assumptions, methods, and model problems that define for a scientific community what the important questions are and how to go about answering them. (Newtonian optics and Freudian psychoanalysis are good examples.) Kuhn's studies revealed two things: that paradigms are tenacious and that one topples another with a quick stroke rather than with small blows. Scientific progress resembles less organic growth than a series of conversions—Eureka!

The value of a paradigm is that it focuses research. Without one, different investigators accumulate different mounds of almost random data, and everyone is too busy trying to make sense of the chaos and fighting off competing theories to make steady advances. The problem with paradigms is that they tend to become inbred and rigid. New advances are increasingly esoteric and inaccessible to all but professionals. Scientists who have something to offer but reject the paradigm are often dismissed as "cranks." Potentially fruitful avenues of research are blocked because they don't arise from accepted premises. Every paradigm, while enabling insights, is also a kind of blindness: It disposes us to see some things and miss others entirely.

Nonetheless, paradigms must shift when old models are conclusively defied by new evidence, as for example when Galileo's discovery that Jupiter has moons helped bring down Ptolemaic astronomy. (Of course, many, including the Church, clung desperately to the old paradigm.) Kuhn's central point is that paradigm shifts, being sudden and disruptive, defy the ideal picture of science as gradual and steady progress toward Truth. So long as a paradigm holds good—so long as a scientific community accepts it and so long as it fits nature reasonably well—research and discovery will be gradual and cumulative. But novelties (unanticipated observations and anomalies) are not readily and happily assimilated by paradigms, at least not for long. Scientific revolutions—paradigm shifts—are inevitable, and necessary, so long as reigning theories are incomplete or blind.

What makes this fact interesting to everyone, not just to scientists, is that a scientific paradigm shift often entails a new, perhaps frightening, worldview. The Copernican revolution displaced man from the center of the universe and forced him to see creation, and his place in it, in a new light. Kepler, Newton, and their cohorts devised a mechanical universe that ran like a watch—one God never had to rewind. Einstein's relativity and Heisenberg's uncertainty, though highly technical in their details, have seeped into the popular consciousness, and the world appears chancier and more random than ever. The most frightening part of all is that the next paradigm can never be foreseen, for we always see the future through the paradigm we have.

From Here to Uncertainty: Modern Physics

Relativity

Relativity does not equal $E = mc^2$. Albert Einstein's signature formula is only a bonus of his larger theory about how things appear from different viewpoints. Of course, if it were just that simple you wouldn't be reading about it here.

Einstein (1879–1955) actually came up with two theories of relativity, called "special" and "general." (The general includes the special as a particular case.) His most interesting conclusions were these:

- Time and space are not absolute fixed quantities. They appear different to people moving at different speeds, though if the difference in speeds is small the difference in appearance is infinitesimal. This is the essence of the special theory.
- If you can't tell the difference between two physical forces or events, then there's no effective difference between them. For example, according to the general theory of relativity, there is no way to tell the difference between an accelerating force and a gravitational force. Thus there is no real difference between acceleration and gravity.

That's mostly it. Of course, the theories came packaged with a lot of mathematics, which makes physicists happy, but in essence Einstein's fame rests on his insight into the subjective quality of observations and measurements. I'll skip the formulae and try to clarify the insight.

Einstein began with an idea that traces back to Newton: Motion is relative. More precisely, the laws of physics look the same whether one is standing still or moving at constant speed. Say you're flying in an airplane cruising steadily at 30,000 feet. You won't actually *feel* the plane's motion unless something disturbs it. If you get up out of your seat to find a copy of *People,* you will exert the same effort, and seem to move the same distance, as if the plane were sitting at the gate. In fact, if you didn't know better, you could be forgiven for thinking as you gaze out the window that it's the clouds and the ground that are moving, not the plane.

Of course, the clouds and the earth *are* moving—just at different speeds from the plane. Which raises an interesting question: How fast is the plane *really* traveling? Let's say that, relative to the ground, the plane's speed is 500 miles per hour. But the ground itself is moving, by virtue of the earth's rotation, at about 1000 mph. To somebody sitting in a space station at rest with respect to the sun, it would look like you are traveling at 1500 mph if the plane is flying east, or, if it travels west, that you're actually flying *backward* at 500 mph (while the earth spins backward at twice that speed).

The upshot of all this is that what looks like 500 mph in one frame of reference will look like 1500 mph or −500 mph, or 0 from another. That is, there is no such thing as "absolute" velocity, just relative velocity. Nonetheless, relativity says that if you are traveling at a *constant* speed with respect to some other frame (we'll call it a "system" from here on out), the laws of physics will look and act just the same to you as it would in that other frame ("system"). Assuming for the time being the unreal situation in which your plane was flying at a fixed speed through a vacuum, it would be just as easy to dribble a basketball in the aisle as it would be to dribble it in your living room, even though, from the point of view of the ground, the floor of the plane is moving underneath the ball at 500 mph. This is "Newtonian" relativity.

The Special Theory of Relativity

What Newton didn't know, and what complicates the picture enormously, is that no matter what system you're in, and no matter how fast it's traveling with respect to another, the speed of light is fixed and constant, if we disallow for gravity. A beam of light passing through your airplane will seem to travel at exactly the same speed to you, to the observer in the space station, and to anyone watching with binoculars from the ground. This speed—approximately 186,000 miles per second—is conventionally denoted c.

Let's bring this discussion down to earth. Suppose you're riding a train doing 80 mph. Let's also suppose for the sake of argument that the train is transparent. If you walk toward the front of the train at what seems to you like 3 mph, to the observer by the tracks you'll seem to be traveling 83 mph with respect to the ground. At least, that's what common sense and Newtonian relativity say.

Now imagine that this observer flashes a strobe up the tracks straight in the direction the train is traveling. The light from the strobe passes through the train; at the same time, it's also moving parallel to the ground. According to physical laws, if the train conductor were to measure the speed of the light as it passes through the train, he would have to come up with c. Likewise, our observer on the ground, measuring the light's speed from his perspective, would also necessarily arrive at c.

There's a major problem here: How could something pass through a moving train at the same speed with respect to the train as with respect to the ground? Think of our earlier example: If you're walking down the aisle at a speed of 3 mph with respect to the train, you've got to be moving faster with respect to the ground—namely, 83 mph (or so Newton would say). Why should the case be different for light? Why doesn't it appear from the ground to move through the train at $c + 80$ mph? Or, if it must travel at c with respect to the ground, why then doesn't it appear to someone in the train to move at $c - 80$ mph?

Here's where relativity comes in. As it turns out, if you're walking through the train at 3 mph, you will *not* appear to someone on the ground to be moving at 83 mph, but rather at a slightly lesser speed. In this case, $80 + 3$ *doesn't* equal 83. The only time the addition will work is if you're standing still on the train: In that case, you will appear to someone on the ground to be moving at $80 + 0$ mph. At the opposite extreme, light moving through the train at c will appear to someone on the ground to be moving at c; the 80 mph doesn't figure at all. But for any motion at a velocity between 0 and c, there will appear to our observer on the ground to be a *contraction* of space and a *slowing* of time.

This conclusion is necessary if we are to solve the problem created by the absolute speed of light. If time and space are different on the train than they are on the ground, then we can have our cake and eat it too. Since speed is distance traveled divided by the time it takes, you can measure your motion on the train as 3 mph and someone on the ground can measure it at less than 3 mph only if your miles and hours are different from his.

This was exactly Einstein's main conclusion when he formulated the special theory of relativity in a paper of 1905. He adapted equations for compression of time and space developed by the Dutch physicist Hendrik Lorentz, who studied electromagnetism, and applied them to all events in space and time. Or rather, he applied them to events that are viewed from two systems moving in uniform, constant, linear motion—a very special case indeed, and one rarely found in experience.

The special theory grew out of Newtonian relativity—the idea that physical laws look and feel exactly the same in two systems in constant relative motion (like a train moving a fixed speed over the ground). But Einstein abandoned one of the seemingly soundest laws of nature: that space and time are absolute—that a mile is a mile is a mile and a second a second a second, no matter what system you're in, whether at rest or moving at 186,000 miles a second.

In fact, what seems like a mile on a moving train will look shorter than a mile from the ground; and what seems like a second on the train will look longer than a second from the station. An odd fact about this is that it also works in reverse: What looks like a mile from the ground will look shorter from the train, etc. Otherwise, you'd know that the train was moving, and not the ground, which contradicts the theory.

Now, Einstein wasn't saying that objects in motion literally contract—that, say, a foot-long hot dog is shorter than twelve inches on a moving train. This is impossible, because relativity is reversible: It's just as good to say the ground is moving as to say the train is moving, so there's nowhere to stand and say, "Here, a foot-long hot dog is *absolutely* a foot long." Einstein was talking about apparent contraction, which amounts to a disagreement between two parties in relative motion as to whose measurements are right.

The sad fact is, nobody's right. To measure something takes time, and it also takes seeing the thing, which requires light, which itself takes time to travel. *Where* you are when you see something—when you receive information about it—determines *when* you think it happened. The point of special relativity is that nobody's in any position to ever say, "This event happened at this definite time in this definite place."

The General Theory of Relativity

We've already fried a whole lot of fish, but Einstein had even bigger ambitions. In his special theory, he showed that physical laws hold across systems moving in uniform motion so long as we abandon the notions of fixed time and fixed distance. In 1916 he took relativity even further, with his "general theory," extending the special theory to include all systems whatsoever, even if they move erratically, elliptically, or with changing velocity with respect to a chosen reference point. He did this by proving that there's no real basis for distinguishing acceleration from gravity—they feel and act exactly the same (which is why, when you ride a roller coaster, changes in speed feel like gaining or losing weight).

Many subtle arguments later—and you'll forgive me for neglecting them here—Einstein showed that the laws of physics could be derived within any system whatsoever, whatever its state of motion, and that not only are distance and time relative, but so are acceleration and gravity, and thus every other quantity that depends on them (such as force and momentum). And there's no way to find, and no basis for choosing, one frame of reference that would yield "true" values of a mile, a second, or a pound.

Among the many revolutionary consequences of the general theory is the realization that time is not independent of space—indeed, that time looks and acts just like a spatial dimension, and that it can be "warped" by gravitational fields. Thus Einstein spoke not of events happening in space at a particular (and unrelated) point in time, but rather of events in a four-dimensional "space-time continuum." This continuum is warped and curved by gravitation; it defies the neat laws of Euclidean and Cartesian geometry, which assume a homogeneity of space and time and a rectilinear universe. (One must use so-called "non-euclidean geometry" to deal with space-time phenomena; Einstein himself preferred the geometry developed by Gauss.)

Even more radically, Einstein questioned the very notions of "space" and "time," which he saw as psychological effects rather than as "realities" of nature. Since the shape of what we call "space" ("space-time" to Einstein) depends on gravitation, which requires material bodies, Einstein decided that space and time without matter are meaningless. Once, when asked to explain the meaning of relativity, Einstein replied, "It was formerly believed that if all material things disappeared out of the universe, time and space would be left. According to the relativity theory, however, time and space disappear together with the things."

$E = mc^2$

And now for $E = mc^2$, which most people equate with relativity. Again, it's only a corollary of the special theory, one Einstein himself didn't make a whole lot of fuss over.

The gory details may be found in an APPENDIX (see p. 239), but here's the quick version: The special theory of relativity says that physical laws must look the same to two observers in uniform constant motion. Among these laws is Newton's conservation of momentum [see p. 85]. But since momentum is mass times velocity, and since velocity will look different to the two observers, relativity forces us to conclude that faster-moving bodies must appear to have more mass. (Take my word on this.) As an object moves faster, therefore, it must appear to gain mass.

Now, if mass is relative to velocity, then what happens to an object when we add energy to it depends on how fast it's already moving. Energy, in the form of force, increases an object's momentum; according to Newton's mechanics, this means simply that we increase its velocity, since mass is assumed to be fixed. But Einstein showed that mass is never absolutely given, but is relative and apparently increases with velocity. So adding energy to an object for all intents and purposes increases its mass.

In fact, if the object is already moving at close to the speed of light, we can hardly affect its velocity at all, and extra energy goes almost completely into increasing its mass. So energy, E, must be to some degree translatable into mass, m. This must be as true for an object moving at the speed of light as for anything else, and we can never make such an object go faster; for such objects, extra energy is the *same* as extra mass. But according to relativity, to say this object is moving at the speed of light is exactly the same as saying that it is at rest while *we* move at the speed of light. In other words, no matter *how* fast something is moving, its mass can be translated into energy.

The proportional factor turns out to be the speed of light squared, giving Einstein's famous formula, $E = mc^2$. What it means is that, if we do not increase the speed of an object, we can add exactly E/c^2 more mass to it by pumping in the energy E. (E has to be pretty huge to make a noticeable difference, though.) And, inversely, we must be able to translate mass back into energy, by the huge factor c^2. The factor is so huge that annihilating a few atoms gives off a tremendous amount of energy, with the effect that whole metropolises can either be destroyed or run off nuclear power plants. Einstein didn't exactly intend either result; he was only interested in the right equations. Technology took the ball and ran with it.

A "Quantum Leap"

Along with Einstein's papers on relativity, the theory of quantum mechanics helped put an end to the days when physics got along with common sense. The Newtonian idea that the smallest particles of matter must behave like the largest, and the belief that theories of the microscopic world will easily correspond with our vision of the world at large, were dead.

Quantum mechanics is the science of how subatomic particles travel, orbit, and leap. (Particularly important is how such activity produces light.) The basic ideas, though mind-boggling, arose out of simple experiments, conducted by Max Planck around the turn of the century, involving radiation of light in various frequencies (colors) by hot black objects.

Planck arrived at strange results. Up until his experiments, physicists assumed that light was a wave form of energy, just like sound. Batteries of experiments backed up this assumption, since light produced interference patterns that only waves could produce. Planck's numbers, however, could only be explained if light emanated not in continuous waves, but in small bursts of particlelike "packets," which he called "quanta" (singular "quantum").

If light really does come in "quanta," how are they produced? Since light is energy given off by matter, its propagation must ultimately trace to the release of energy at the atomic level. The mechanics of this are still debated, but the central theory is that of Niels Bohr, a Danish physicist who applied Planck's quantum theory to the atomic model developed in the 1910s.

According to this model, every atom is like a miniature solar system, with the nucleus in place of the sun and electrons in place of planets orbiting the nucleus. According to Bohr, electrons revolve around the atomic nucleus along particular and fixed orbits. If we bombard an atom with energy we may "excite" its electrons to jump from one orbit to another, but they can never be made to reside somewhere in between—in fact, they can't even be said to "exist" in between: They disappear from one orbit and appear at

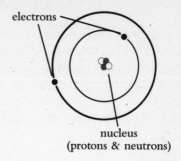

electrons

nucleus
(protons & neutrons)

another. And once we remove such external stimuli, the electrons will "jump" back to their original orbits, releasing energy in the process.

This is the famous "quantum leap," first described by Bohr in 1913: When an electron jumps from an outer orbit to an inner one, energy is released in the form of a quantum of light (called a "photon"). The abruptness of the change in energy, and the fact that electrons jump instantaneously from one position to another (nonadjacent) one without actually passing physically in between, explains the common use of "quantum leap" to mean "radical and sudden change of circumstances." (Physicists prefer the term "quantum jump" when describing the theory.)

The problem with Bohr's theory is that, while it does account for a variety of observed phenomena, it can never be proved through observation. You can't put an atom under a microscope and watch its electrons jumping. And in fact a number of rival methods and theories sprang up in response to Bohr's research, most compellingly at the Copenhagen Institute for Theoretical Physics, where Werner Heisenberg did some of his most important work, and where Erwin Schrödinger proposed a radical reinterpretation of the atomic model. The story continues in "The Uncertainty Principle."

The Uncertainty Principle

From uncertainty it arose, and to uncertainty it has returned. What Werner Heisenberg's "uncertainty principle" really means kind of depends on whom you ask. Ask a hundred people and you'll get 60 blank stares, 30 shrugs, and 10 versions of the answer "We change the world by observing it," which isn't quite true. The irony is that Heisenberg hoped to *reduce* the confusion bred by modern physical theories—quantum mechanics in particular.

What the principle basically says is that there is no way to precisely measure most essential properties of subatomic behavior. Or rather, the more precisely you measure one property—say, the momentum of an electron—the less precisely you can know another—in this case, its position. The more certain the one, the more uncertain the other.

Heisenberg discovered this distressing fact in attempting to deal with vying theories of light. According to Niels Bohr's quantum theory, which Heisenberg preferred, light is emitted discontinuously by atoms in packets when electrons make a "quantum leap." According to others, such as Erwin Schrödinger (of cat fame), quantum theory fails because it can't account for the ways in which light behaves like a wave.

Heisenberg himself was dissatisfied with Bohr's theory, because it relied on a picture of the atom that could never be proved. But he thought Schrödinger's rival picture was more wrong, and to prove it he set about examining more closely what we *can* say for sure about electrons. In the process, he scrutinized the common measurements—position, velocity, momentum, energy, and time—physicists used in propounding their theories. By 1927, he had come to a surprising conclusion: that both the quantum theory and the rival wave theory, as they were then formulated, were fraught with insuperable uncertainties.

Heisenberg began by thinking hard about the very process of scientific observation, which may generally be reliable when dealing with everyday objects, but which runs into grave difficulties when it comes to subatomic particles. His first point was this: You can't observe the position of an electron except by bouncing something—light, for example—off of it. In other words, you have to introduce a form of radiation, which has energy of its own, and this energy will disturb the path of the electron to a greater or lesser degree.

In fact, the more precisely you wish to locate the electron, the more you have to disturb its velocity (and thus its momentum), because you have to add more energy. Conversely, if you wish to precisely measure the electron's momentum (expressed in its velocity), you must minimize the interference of the radiation. But by doing so, you make it impossible to precisely locate the electron's position.

To sum up, radiation of high energy will give you more precise data about where the electron happens to be at a given time, while destroying the evidence of its initial velocity. Radiation of low energy will give you more precise data about how fast the electron is moving at a given time, while fuzzing up the data as to where it is. Even stranger, the very act of observing an electron's position will make it "behave" more like a particle, while the act of measuring its energy will make it "behave" more like a wave.

Heisenberg came up with an interesting little formula to express these frustrating facts, the gist of which is that if you multiply uncertainty of position by uncertainty of momentum, the product can never be smaller than a certain positive number called "Planck's constant." That is, the uncertainty can never reduce to zero, and the better you measure one quantity the more uncertain the other.

The point is not that our knowledge of atomic particles is uncertain because our measuring techniques aren't good enough.

The point is rather that no technique whatsoever can ever overcome a fundamental uncertainty or "fuzziness" in quantum behavior. Electrons may in fact behave like precise points moving at precise speeds, but we'll never be able to know; it's just as likely that they don't, so propositions to either effect are meaningless and useless.

In practical terms, what the uncertainty principle suggests is that you cannot treat particles or quanta as if they were like the objects we encounter in daily life—objects we can put our finger on and say, "*Here* is this object now, and *there* is where it's going." The essential aspects of a particle (position, velocity, momentum, energy) can never be precisely observed at once—the act of observation itself inevitably and irretrievably distorts at least one of those quantities. The best we can hope for is to take measurements and make predictions that are *probable* or statistical.

Such seemingly defeatist notions got up the dander of several formidable physicists, the most famous of whom was Albert Einstein. His rebuttal follows.

"God Does Not Play Dice"

"At any rate, I am convinced that *He* does not play dice."

Albert Einstein, letter to Max Born, 1926

Even the most radical and innovative thinkers, after pioneering new systems of thought, never completely break with the old. Sigmund Freud remained in essence a nineteenth-century scientist, and so, in many ways, did Albert Einstein.

Einstein, by helping formulate the dual particle/wave model of subatomic entities, contributed as much as anyone to the emergence of quantum mechanics. But in the end he was unable to accept its conclusions. When he remarked, in a letter to his colleague Max Born, that "*He* [God] does not play dice," Einstein was repudiating the uncertainty principle and all other claims that chance plays an essential role in physical events. He believed that the universe is lawful and orderly; whatever God is, even if He is just a metaphor for the unfolding of space and time, He is a good Newtonian.

The essence of Newtonian physics is physical determinism. Given a complete description of a situation or system—its objects, their masses, total energy, etc.—you should in principle be able to precisely predict how that situation would change in any given amount of time. For example, if you know how fast a ball is pitched to a batter, how much energy that batter puts into his swing, at what position and time he hits the ball, and how the wind is blowing, you ought to be able to predict exactly where he's going to hit it.

But Werner Heisenberg—with whom Born agreed—dealt a serious blow to the Newtonian worldview. The logical conclusion of quantum mechanics, Heisenberg said, is that cause and effect, interpreted strictly, are empty notions. Heisenberg showed that, at the subatomic level at least, you can't ever know all the initial conditions of a situation—at best, you're dealing with probabilities and statistics. Therefore, atomic behavior is indeterminate: It cannot be predicted. Taking this a step further, Heisenberg repudiated the classical notion of causality; in his 1927 paper announcing the uncertainty principle, he branded as "useless and meaningless" the assumption that "behind the statistical universe of perception there lies hidden a 'real' world ruled by causality." Since we can never perceive or measure subatomic conditions, and since we can't even know if causality applies, the notion is best abandoned.

This idea repelled Einstein, who shouldn't be mistaken for a relativist simply because he formulated relativity. Einstein did destroy the belief that there is any absolute reference point for physical measurements; still, when it came to relationships among relativistic systems, he offered precise formulas yielding hard numbers. Einstein's universe has a definite shape and though relativistic is continuous and predictable. In short, he couldn't imagine a world that could only be described in the fuzzy expressions of statistics.

Einstein strove mightily for years to convince partisans of quantum theory that their assumptions must be wrong, and that chance plays no part in physical events. His poignant losing battle was inspired partly by the fact that his own work relied on a continuous, causal universe—the space-time "continuum"—denied by Heisenberg's interpretation of quantum theory. And behind that work is a deep sense of universal order and continuity.

These qualities, rather than an omnipotent personal deity, are what Einstein understands as God. (Please note that he did not write "God does not play dice with the universe" but "*He* does not play dice.") "I do not believe in a personal God," Einstein said, "and I have never denied this but have expressed it clearly. If something is in me which can be called religious then it is the unbounded admiration for the structure of the world so far as our science can reveal it."

What Einstein could not imagine is a universe in which basic building blocks such as electrons roamed free, lawless and unfettered by causality. Such a universe could have no overall design or coherence. Whether or not Einstein was right remains, along with the more abstruse claims of quantum mechanics, in dispute to this day.

Is Your Logic Fuzzy?:
The New Mathematics

Russell's Paradox (Metalanguage)

A man comes up to you on the street and says, "Whatever I tell you is a lie." Is he telling the truth? This is a version of the "Cretan liar paradox," so-called because it traces to the wit Epimenides, from Crete, who said that "all Cretans are liars."

Nobody knew what to do with such head-spinning self-consuming riddles until British philosopher Bertrand Russell (1872–1970) faced a similar paradox in logic. Russell's version arose in his attempt to deal with the discovery that mathematical "truth" isn't what we once thought it was.

Centuries earlier, Euclid had put forth five axioms of geometry, and everyone accepted them because they seemed applicable to reality. But in the nineteenth century it was shown that equally valid and consistent geometries could be built from seemingly "false" assumptions. Euclid held, for instance, that given a straight line and a point not on that line, only one other line could be drawn through the point that would be parallel to the first. Such an assumption seems intuitively true. But if we posit contrarily that *more* than one parallel line can be drawn through that point, or on the other hand that *no* parallel line can be drawn, we can still derive a rigorous geometry. (Such geometries are called "non-euclidean.") These premises are no more "true" or "false" than Euclid's, at least by any hard standard of proof; indeed, Einstein's theories of relativity *require* non-euclidean geometry.

As with geometry, so with arithmetic and all other branches of math. If the validity of mathematics is to be assured, then, we have to look elsewhere than to intuition, common sense, or practical experience. To place the doubtful concepts of math on a surer foundation, Russell thought, one need only find the simple components from which they are constructed. These lay in logic, whose principles are about as certain as anything gets. What Russell wished to do, therefore, was analyze arithmetic down into universally accepted logical notions, and then rebuild math back on top of those notions. (Most math can be derived from pure arithmetic.) Along with mathematician Alfred North Whitehead, he devoted the three volumes of *Principia Mathematica* (1910–1913) to this enterprise.

The first stumbling blocks were the three undefined terms of arithmetic: "zero," "number," and "successor" (as in the statement, "the number 1 is the successor of zero"). Every other proposition of arithmetic follows once these three terms are defined. To this end, Russell turned to the logic of "classes" or sets, which was on the leading edge of theoretical math. Russell thought he could define both "zero" and "number" with the logical concept of a set of sets, or a "class of classes." But as it turns out this concept isn't at all well behaved; it leads to contradictions and paradoxes, in fact to a kind of logical Cretan liar paradox.

The problems begin when we attempt to treat classes in the same way as the objects they contain. In most cases the difference is obvious. Think of a six-pack of beer as a set or "class" of beer bottles; obviously the six-pack isn't itself a bottle, and a six-pack can't contain another six-pack. However, a case of beer may contain four six-packs, making it a "class of classes" (a set of four six-packs, which are themselves sets). The question is whether there's less of a difference between a case and a six-pack than there is between a six-pack and a beer bottle. After all, both a case and a six-pack are classes, so they probably have similar properties.

Here's where the paradox comes in. A set of physical things cannot contain itself, since a set is not a physical thing. For example, a box of bottles doesn't contain itself, since a box isn't a bottle. The same can even be said of certain sets of sets. Take for example the class of ethnic groups in California. Each group is itself a set—the set of Latinos, the set of Chinese, the set of African-Americans, the set of European-Americans, etc. But the set of ethnic groups isn't itself an ethnic group, so it doesn't contain itself.

Likewise, the set of all cats doesn't contain itself, since it isn't a cat. But how about the set of all non-cats? Either a thing is a cat or it's not, and a set is not a cat: Therefore, the set of all non-cats *must contain itself*. Here's an even simpler example: the set of all sets. Since the set of all sets is a set, too, it must also contain itself. Here's where the fun begins.

Since a set either contains itself or it doesn't, we may divide all possible sets into two groups or classes: the class of sets that don't contain themselves (call this N, for "no") and the class of sets that do (call this Y, for "yes"). The set of beer bottles belongs in N, as do the sets of cats and of ethnic groups in California. Sets such as the set of all sets and the set of non-cats go into Y.

A man comes up to you on the street and says, "The set N contains itself." Do you believe him?

This is Russell's paradox, which doomed any attempt to base arithmetic on the logic of sets. The answer, if you think about it, is that there's no answer: We've arrived at a logical breakdown. For if N contains itself, then N is by definition a set that doesn't contain itself. But if we assume that N doesn't contain itself, then it is a set that doesn't contain itself, so it must belong to N after all.

Russell realized that the problem here is treating all sets alike, for this is how we get ourselves tangled up in the issue of sets being able to contain themselves. To get rid of such embarrassments he proposed that sets should be considered according to what he called their "type." A set of plain objects is of the lowest (most basic) type—call it a "type 1" set. Such sets may contain objects only, not other sets. Next up the ladder are "type 2" sets, which may contain objects but also type 1 sets. Your basic set of sets (like our box of bottles) is a type 2 set; it can never contain itself, since it only contains sets of a lower type. Higher up are the type 3 sets, which may contain objects, type 1 sets, and also type 2 sets. Likewise these can never contain themselves. And so forth. Once such distinctions are given, the question of whether a set contains itself becomes meaningless.

Russell thought that a similar logic could clear up paradoxes of language such as the Cretan liar paradox. Though he didn't use the term, what he basically proposed was the concept of "metalanguage," language that makes a statement about itself. We might call the basic type 1 set just a plain "set," and call a type 2 set that includes it a "metaset." Similarly, we can call statements about objects or simple relations, such as "The cat is on the mat," simply "plain language" (or "object language"). Statements that refer to plain language, its meaning, or its truth, however, are not plain language but "metalanguage." (And language about metalanguage is "metametalanguage.") To avoid the Cretan liar paradox, all you have to do is separate metalanguage from language and take care not to treat truths about one as truths about the other. Paradoxes such as "Whatever I say is a lie" are banished to the realm of nonsense, since they attempt to equate language with metalanguage, violating the hierarchy and looping language back on itself.

However, holding language and metalanguage apart is a lot harder than it seems. "The previous sentence has 13 words" is a clear and simple case of metalanguage, which also happens to be true. But try another statement. "This paragraph has eight sentences." The trouble with this metalinguistic statement is that it counts itself as one of those sentences—that is, it refers not just to ordinary language but also to metalanguage. Is it therefore a metametalinguistic statement? And then is that last question a metametametalinguistic one? So would the fact that this paragraph has eight sentences become a metametametametalinguistic statement?

No matter. Russell was concerned only with math and not with ordinary language. But even his improved approach to the logic of mathematics wasn't improved enough, even after he cleaned up set theory and incorporated seemingly more reliable notions from propositional logic. As it turns out, you can't get self-referentiality or circularity out of the system: There's always a Cretan liar paradox lurking in the axioms. The hidden paradox in Whitehead and Russell's *Principia* was brought to light eighteen years later when an Austrian mathematician turned Russell's paradox back against Russell.

Gödel's Incompleteness Theorem

To every ω-consistent recursive class κ of *formulae* there correspond recursive *class-signs* r, such that neither ν Gen r nor Neg (ν Gen r) belongs to Flg (κ) (where ν is the free variable of r).

Kurt Gödel, "On Formally Undecidable Propositions in
Principia Mathematica and Related Systems I" (1931)

If your mind is boggling over that quotation, join the club. Even most mathematicians couldn't fathom Kurt Gödel's response, published in 1931, to Whitehead and Russell's masterwork of symbolic logic, *Principia Mathematica.*

A quick summary of what Gödel was getting at is this: Any complex formal system of thought, such as standard logic or arithmetic, is bound to be incomplete. In slightly more precise terms: Given a finite number of elementary assumptions ("axioms") and rules for deducing propositions from them, you will always, if the axioms are consistent, be able to produce at least one true statement the system cannot prove.

Gödel's larger point was that formal sign systems such as pure arithmetic can never be used to prove their own completeness or consistency. (A "complete" system will generate all true statements; a "consistent" one will not generate contradictions.) Supplementing or extending the system can never remedy this situation; we must look outside it for reassurance. But then we have to show that such outside methods are themselves reliable, which is liable to be even more difficult.

Gödel's "Incompleteness Theorem" and its proof are extremely abstract and involuted, but with lots of shortcuts I can lay out the essentials. If you've ever studied geometry you'll at least be familiar with axioms—that is, the basic assumptions used for deriving other true statements. Euclid founded his geometry on five such axioms, though in the last century or so their self-evidence has been questioned. Examples of axioms in arithmetic include "zero is a number" and "a number is equal to itself." While we take them for granted, such things cannot be proved.

Ever since Aristotle scientists and philosophers have attempted to extend Euclid's success to other fields of knowledge. With a set of axioms and a deductive logic, they hoped, one could accept or reject hypotheses with absolute certainty and ultimately generate all possible truths. But with the rise of the experimental method, natural science gave up this dream, leaving it to pure mathematics. And there were notable successes: in the late nineteenth century Gottlob Frege and Giuseppe Peano developed notation systems that united math and logic, and in the 1910s Whitehead and Russell finally seemed to set arithmetic on the same firm axiomatic basis as Euclidean geometry.

Their triumph, however, was to be short-lived. What Whitehead and Russell hoped to establish was a system, involving a small number of axioms and rules of deduction, that was both *consistent* and *complete*. A system is consistent if contradictory statements cannot both be derived within  it. That is, if you can "prove" (deduce) the "theorem" (formula) "$2 + 2 = 4$," then you can never prove the contradictory theorem, "$2 + 2 \neq 4$."

Whitehead and Russell's initial axioms seemed consistent, given their rules of deduction, and for now we'll take consistency for

Eureka!

granted. We're faced, however, with the question of whether their system is complete—that is, whether it can generate *all* true arithmetic formulas, and whether all true arithmetic formulas are capable of being proved within the system. This much harder question is what occupied Gödel, and the one he answered in his paper "On Formally Undecidable Propositions."

Once again, what Gödel proved was that no finite, consistent set of axioms can ever be complete, and that no matter how many more axioms you add to a formal logical system to make up for some deficiency, it will always be possible to find a true theorem that cannot be proved.

Gödel's proof is a brilliant piece of work. Using the symbolic logic of the *Principia,* he devised a way of pairing each symbol, axiom, proposition, or proof with a unique number, called its "Gödel number." For the sake of argument, let's say that the Gödel number of the axiom "$x = x$" ("a number is equal to itself") is 156. (Actually, even equations that simple have huge Gödel numbers.) By a simple rule of deduction, we can derive from "$x = x$" the theorem "$0 = 0$." Let's say that the Gödel number of "$0 = 0$" is 72. Gödel showed that a statement like "'$0 = 0$' is a true theorem" can itself be reduced to a formula relating Gödel numbers—in this case the numbers 156 and 72.

Now, such a statement is not itself a formula in the system—it is a statement about the system. Other such "meta-arithmetical" statements would include: "'$2 + 2 = 5$' is false," "If the theorem T is true, then the system is inconsistent," and "Theorem S cannot be proved." The way Gödel rigged his numbers, every such meta-arithmetical statement is mirrored by a mathematical relation among Gödel numbers—in other words, meta-arithmetical statements can be modeled or "translated" within arithmetic.

Gödel then pulled a rabbit out of his hat. Using the Gödel numbers, he constructed an arithmetical theorem—we'll call it G—whose meta-arithmetical translation was "the formula G cannot be proved." If G is a true theorem, then "G cannot be proved" is also true, and therefore the system is incomplete—we have found a true theorem that can't be proved within the system. But it turns out that if G is false, what that means is that G cannot be proved (if the system is consistent), and therefore the statement "G cannot be proved" is true. But since G *means* that "G cannot be proved," G is true. This contradicts the premise that G is false, and so the system is inconsistent.

To sew up the proof, Gödel put together a formula that corresponded to the statement, "If arithmetic is consistent, then G can be proved." The formula turns out to be a true theorem in the system. But what Gödel showed is that G can be proved only if its opposite can also be proved; however, if arithmetic is consistent such contradictions never arise. Therefore we can never prove that arithmetic is consistent, at least if we rely on the assumptions embodied in arithmetic.

The crux is that Gödel used arithmetic to prove that the completeness and consistency of arithmetic can't be proved. No finite number of supplementary axioms will "fix" the situation, either. If there are proofs, they lie beyond logic, the axiomatic method, and finally mathematics. And any extramathematical method in its turn would have to demonstrate its consistency, and now we're traveling in circles it's doubtful we can ever escape.

Gödel's Incompleteness Theorem would seem to bode particular ill for artificial intelligence, at least as we know it today. Computers still are, and may forever be, logic machines that operate on finite data and with a finite number of instructions. Program in a billion directives ("axioms" and "rules of deduction"), and still a computer can never arrive at, or even prove, all the truths human minds discover on their own.

The Prisoner's Dilemma
(Game Theory)

You and a criminal accomplice are hauled down to the police station and each put in an isolated room. A prosecutor tells you the police have sufficient evidence to send both of you to jail for a year, but not enough to get you convicted on more serious charges. But if you confess and agree to testify against your partner, you'll get off free for cooperating while he goes to jail for three years. If you both confess to the greater crime, however, the cops won't need your cooperation and each of you is going to serve two years. You're led to believe your partner's been offered the same deal. What do you do?

This is a common version of the "prisoner's dilemma," a famous problem in game theory, the mathematics of decision. (There are other dilemmas in game theory, such as the "chicken dilemma," which we'll get to later.) Perhaps you haven't been arrested lately and wonder why you should care. Actually, one doesn't have to look far to find other prisoner's dilemmas in everyday life. Given the chance, do you cut to the front of a long line? What's your response to those obnoxious public-broadcasting pledge drives? Do you deal with office conflicts by stonewalling or by compromising? In each case, you're faced with a problem similar to that of the prisoners: Are you really best off behaving selfishly?

The dilemma is that a choice cannot be made on purely rational grounds. To see why, let's return to our initial scenario. Looked at one way, you're better off confessing, but looked at another, you're best keeping quiet. Here are the possible outcomes arranged in a matrix:

	Partner keeps quiet	Partner confesses
You keep quiet	1 year for you	3 years for you
	1 year for partner	0 years for partner
You confess	0 years for you	2 years for you
	3 years for partner	2 years for partner

Obviously, as far as you're concerned the best possible outcome is that you confess and your partner doesn't. (In the language of game theory, to look out for yourself no matter what is called "defection.") And even if your partner confesses, you still profit by defecting, since if you remain silent you're looking at three years in the clink, while confession gets you only two. In other words, no matter what your partner does (and you have no way of knowing his decision) you're better off defecting.

But if your partner is as smart as you, he's going to reach the same conclusion: The rational move is to confess. This logic will thus get you both two years in jail. Is this really "rational" when by both keeping mum ("cooperating") you would each only do a year's time? On the whole, mutual cooperation is best, since the combined amount of time you would serve is two years rather than three or four.

So you should cooperate, right? Well, suppose your partner *doesn't* reach this conclusion, or that he does but decides to exploit your trust by defecting. In that case, you're looking at the worst possible outcome: three years making license plates. What will it be: Do you trust him, or not? Is cooperation or defection more rational?

This and similar problems are the provenance of game theory, more or less the invention of mathematician John von Neumann (1903–1957). A Hungarian prodigy who settled in the States, von Neumann helped develop the A-bomb and invent the digital computer, among other achievements. He also loved games of strategy, especially poker and chess, and in the 1920s and 1930s developed a math to describe their structure. Von Neumann did so partly to better understand games, but mostly because he believed game theory could provide a scientific basis for the study of gamelike situations in the larger world. He coined the term "game theory" in *The Theory of Games and Economic Behavior* (1944, with Oskar Morgenstern). Economic behavior is a "game" in von Neumann's broader sense: a situation defined by competing interests and in which everyone is looking out to maximize his profits.

As it turned out, game theory was a bust for economics but useful elsewhere. After World War II von Neumann was hired by the RAND Corporation, where he applied game theory more fruitfully to Cold War strategy. Think back to the 1950s and imagine having to decide whether the United States should build an arsenal of H-bombs. Let's assume the Soviet Union, the "enemy," is perfectly capable of doing the same. Your possible choices are two: Build the arsenal or don't build it. There are four possible outcomes:

1. Neither the U.S. nor the U.S.S.R. builds an arsenal—the status quo is preserved.
2. The U.S. builds an arsenal but the U.S.S.R. doesn't—the U.S. is in a position to potentially destroy the Soviet Union and dominate the world.
3. The U.S.S.R. builds an arsenal but the U.S. doesn't—the Soviets are in a position to potentially destroy the U.S. and dominate the world.
4. Both the U.S. and the U.S.S.R. build arsenals—an arms race escalates, neither side dominates, lots of money is spent, and the entire world now faces the threat of devastating nuclear war.

If you study this "game," you'll notice that it's a kind of prisoner's dilemma. No matter what the U.S.S.R. does, it's in your best interest to build the bombs. (If they don't, you've become the world's supreme power; if they do, you at least keep even with them.) But if the Soviets reach the same conclusion, you're both going to spend tons of money just preserving the balance of power while generating volatile nuclear stockpiles. The ideal outcome is "cooperation": Each side refrains (outcome 1). But do you trust the other side? In the end neither did.

Though von Neumann got game theory going at RAND, it was not he who discovered the prisoner's dilemma nor he who studied its implications. Von Neumann concentrated almost exclusively on what he called "zero-sum games." In such games the total payoffs are fixed, and an opponent's gain is necessarily your loss. Most board games, for example, are zero-sum: If your opponent wins, you lose. Poker, too, is zero-sum: winner takes all.

One of von Neumann's colleagues at RAND, John Nash, extended game theory to cover games between two people that are not zero-sum. His theory was that in such games there is always an "equilibrium point": Given that your opponent won't change his strategy, neither would you. Take this game as an example:

	K chooses heads	K chooses tails
You choose heads	You win $1 K wins $3	You *lose* $1 K wins $4
You choose tails	You win $2 K wins nothing	You win $1 K wins $2

In this game the "equilibrium point" is tails/tails (lower right-hand box). This is because no matter what K does, it's always to your advantage to choose tails, and the same goes for K. And even if K were given the chance to change his strategy, you'd still choose tails, and vice versa.

What Nash didn't realize at first, or accept till later, is that just because an equilibrium point exists, it doesn't mean that in real-life games people will choose it. This is especially so in the case of "iterated" games—games between two or more players that are repeated over and over, with the same fixed strategies and pay-offs. Let's look again at the prisoner's dilemma, which was in essence discovered by two other RAND scientists, Merrill Flood and Melvin Dresher, in 1950. (They discovered the game's form; actual prisoners were introduced, and the dilemma named, later that year by Albert Tucker.) The equilibrium point is mutual de-fection: Given that your partner/opponent has chosen a strategy and that it can't be changed, you're always better off defecting.

But suppose you and an opponent play a prisoner's-dilemma type game a hundred times in a row. Let's say these are the payoffs:

	K cooperates	K defects
You cooperate	$2 for you $3 for K	$0 for you $4 for K
You defect	$3 for you $1 for K	$1 for you $2 for K

No matter what K does, you're always better off defecting—you always win a dollar more. The same goes for K: No matter what you do, she wins a dollar more by defecting. But mutual cooper-ation is better for both of you than mutual defection; the worst scenario for you is to cooperate while K defects.

If this game were a one-shot deal, and you and K couldn't strategize beforehand, then the logical thing to do is defect, since you don't know K's strategy and can't change it. But things are very different in a repeated game. Let's say K decides to cooper-ate, hoping you will too, ensuring the best mutual outcome. You, on the other hand, follow the one-game logic and defect. You win big ($3) while K wins her smallest possible sum ($1), and so the next time around K decides to "punish" you by defecting herself. In essence, by defecting K is depriving you of two dol-lars—twice the extra profit you gained by defecting the first time.

So while defection is safe, you could potentially win a lot more money if both you and K cooperated. Of course, if K cooperated while you defected every single round, you'd end up with the maximum take of $300. But if K is rational she'll follow suit by defecting every time, earning herself $100 more than she would by cooperating every time. What, then, is the best strategy?

Game theory, with the help of computer models, has the answer: It's called "tit for tat." You begin by cooperating. If K also cooperates, you cooperate again in round two. You continue like this until K defects, at which point you "punish" her by defecting the next round. The reason this strategy works is that you're using the game to send K a message: "I will always do whatever you did the last round; and since you never benefit from my defection, you should thus always cooperate with me, ensuring the best mutual take." In other words, you're inviting her to join you in playing against the *game itself* rather than playing against each other.

In real life, tit-for-tat means treating other people the way they treat you, but always acting nice at the beginning. Cutting to the front of a line may be best for you, but it's bad for everyone else, and if they responded in kind chaos and fistfights would ensue. Likewise, everyone benefits if you give to public television; you could freeload, but if everyone else followed suit, there'd be no more *Masterpiece Theatre*. Of course, it would be silly for you to cooperate if nobody else did; but since everyone realizes this, and nobody likes chaos, most people do cooperate.

Another game-theory dilemma we meet in real life is called "chicken" (a name coined by Bertrand Russell, believe it or not). You and a friend hop on bicycles and speed toward the edge of a cliff. The first to stop or change course is the "chicken." If both of you stop ("cooperate") simultaneously, then nobody's chicken, but neither does anyone win. The best outcome for you is if your friend stops first: That way you win, and he's chicken. The worst outcome for everybody is if neither of you stops at all—that is, if both of you "defect": You both go over the cliff. What do you do? (This game differs from the prisoner's dilemma in that mutual defection is *worst* for both parties.)

As you may have noticed, game theory—while mathematically rigorous—hasn't yet solved all human conflict. In the first place, for the theory to work it must be clear who all the players are, and the payoffs must be expressible in numbers (or at least probabilities). This isn't always the case in the complicated games of society or politics. Second, what constitutes "cooperation" or "defection" may be rather fuzzy—there's a lot of middle ground in real life, and opponents tend to disagree on the terms (what seems sufficient to one party may not satisfy the other). Yet it's better to have tools than not to have them, and game theory is a remarkably interesting tool with real applications in physics, ethics, engineering, and even biology. The evolution of a species, for example, can be understood in terms of game theory, but that's another long story.

Fuzzy Logic

The axioms and laws of mathematics are very good at doing certain jobs. We know with absolute certainty, for example, that $2 + 2 = 4$, and that the angles of a triangle add up to $180°$—these follow necessarily from axioms. Math is also useful when applied to fixed physical quantities. Einstein used math to show that nothing can travel faster than the speed of light, which is a fixed quantity. Gamblers like Blaise Pascal invented statistics to calculate the probabilities of clear-cut outcomes—say, that a die will come up showing four. The weatherperson on TV uses numbers to predict the chance that it will or won't rain tomorrow.

You might call all these calculations the product of "hard logic," which ultimately owes its methodology to Aristotle. Recently, however, a small posse of engineers and physicists have put hard logic aside in favor of what they call "fuzzy logic," the science of indeterminate quantities. It's fine, according to fuzzy logicians, to say that there's a sixty percent chance of rain, so long as you can define what counts as "rain." The weatherperson assumes that there are two options: it either rains or it doesn't.

But in fact the concept "rain" is fuzzy. If two drops of water fall from the sky, is that "rain"? How about fifty drops? A thousand? Suppose the fog is thick and low, and you feel drops of water on your face. Is that rain? Where, the fuzzy logicians ask, do you draw the line? When does not-rain become rain?

If these sound like Zen riddles to you, you're not alone. In fact, proponents of fuzzy thinking such as USC professor Bart Kosko hype their new science as an East-meets-West synthesis. And while fuzzy logic is more scorned than celebrated in the States, it's all the rage of Japanese industry. You may have heard of the new "smart machines" coming out of Japan: smart washing machines, smart soda-vending machines, smart microwaves, smart camcorders. Such devices are programmed to deal with states between "on" and "off," quantities more finely gauged than "high," "medium," or "low," answers between "yes" and "no."

If "fuzzy logic" has an origin, it lies in logic's attempt to adapt to Russell's paradoxes and Heisenberg's uncertainty. Polish logician Jan Lukasiewicz developed a "multivalent" logic in the 1920s that refined the binary, yes-no logic of Newtonian physics to allow for indeterminate states. In 1965, Berkeley mathematician Lotfi Zadeh applied this new logic to the theory of sets in his paper "Fuzzy Sets," which then lent its name to the logic.

The sets you learned about in grade school were well defined. Either a thing was in a set or it was not. The number 2 is in the set of even numbers and not in the set of odd numbers, and the two sets have an "empty intersection"—which is to say, no number is both even and odd. (By convention the number 0 is neither.) Zadeh's fuzzy sets, however, are, well, fuzzy. Some things may belong and others not belong to such sets, but there is a third class of things that belong *to a certain degree.*

The set of even numbers and the set of odd numbers are hard sets. The set of men and the set of women are virtually hard—there's a slight fuzziness about hermaphrodites and transsexuals. But how

about the set of tall people? No one would call a 4' 2" man "tall" but everyone would call a 7' 6" woman so. But where do you draw the line? Does a man 5' 10" tall belong in the set of tall people, or not? Would an Asian agree with a European, or an Italian with a Swede? Height is both subjective and continuous, so that it is impossible to set a fixed height that would count as tall and automatically exclude everyone beneath it. If 6' is tall, how about 5' 11.99"? Once you start thinking about such questions, about matters of degree, your thinking is getting fuzzy.

To take another example, the set of happy people is fuzzy, because most of us are happy to a degree—maybe to a greater degree, maybe to a lesser, but almost never absolutely happy or absolutely miserable. An opinion poll asking such questions as "Are you happy with the president's performance?" is flawed, since most people are happy or unhappy only to a degree. Adding a scale of 1 to 10 only partly helps, since we still have a range of hard numbers applied to a continuum of opinion. Not all "5s" are going to be equal.

Fuzzy sets are the key to fuzzy machines. Most of the devices we're familiar with are "dumb"—that is, rigidly programmed. Your television is either off or on; the brightness is set to 6 and the contrast to 3. A thermostat-controlled heating system (which we will meet again in CYBERNETICS, p. 134) is the classic dumb machine. When the temperature falls below a set temperature, the heat switches on; when it rises above that temperature, the heat switches off. The mechanism is binary: The heat is either "on" or "off," and when it's on it's always on to the same degree.

Fuzzy machines, on the other hand, use fuzzy sets to produce more flexible responses. Thermostats "think" it is either hot or cold, and they send instructions to turn off or on in response; fuzzy instructions allow it to be hot or cold to a degree. If we decide that 65° is the perfect temperature, we can tell a "smart" heater/air conditioner to modulate its behavior depending on how much the actual temperature differs from 65°. The unit would never be just on or off—it would always be on to a variable degree, mixing and matching instructions.

The famous fuzzy washing machine works on the same principles, keeping an electronic eye on a range of variables, calculating weighted averages, and adjusting its instructions in response. What kinds of fabrics do we have here? How dirty are they? Are we dealing with grease, catsup, coffee, dirt, sweat? How big is the load? All these quantities come in degrees, and the smart washing machine calibrates its responses interactively. In similar ways, smart camcorders precisely adjust focus and aperture; smart TVs monitor and adjust the brightness and contrast of a shifting image.

According to fuzzy logicians, the entire world of facts—like tallness or TV images—is fuzzy. David Hume divided statements into "relations of ideas" and "matters of fact" [see HUME'S FORK, p. 30]; the former are necessarily true, while the latter might be true or false. To the fuzzy way of thinking, *nothing* empirical is either absolutely true or absolutely false, just true or false to a degree. Modern scientists acknowledge that their theories and deductions are never absolutely certain, just highly probable. The fuzzy response to this is that probability still relies on indefensible assumptions, such as that a particle is or isn't where probability predicts it will be. Scientists don't say that a particle is 70 percent there and 30 percent not there, just that there is a 70 percent chance that it is 100 percent there. That's not fuzzy.

What is fuzzy is this: The world is gray. Nothing is pure black and nothing pure white. When we apply black-and-white reasoning to a gray world, we must treat something true to a degree (say, that a glass is somewhat full) as either wholly true (the glass is full) or wholly false (the glass is empty). Each step in a reasoning process requires such a simplification and therefore adds another layer of arbitrariness and error. The more reasoning you throw at something, the *farther* you get from the actual case, not the closer.

What all this adds up to depends on whom you ask. At the very least, fuzzy logic does build better machines; the question is whether it really amounts to a mathematical revolution. Fuzzy proponents vigorously toot their own horns, which can be annoying, especially since fuzzy logic still appeals to standard geometry and algebra, and since fuzzy machines still use computer chips that process binary digital data. For these and other reasons, most Western mathematicians and engineers consider *fuzzy* just a nineties buzzword, old wine in a new bottle. But their tune might change once Japan finally buries the U.S. economy.

Big Bang to Big Mess: Entropy, Chaos, and Other Reasons the Universe Is Going to Hell

Entropy

I propose to call the magnitude S [energy unavailable for work] the *entropy* of the body, from the Greek word [*tropē*], transformation.... The energy of the universe is constant—the entropy of the universe tends toward a maximum.

Rudolph J. E. Clausius, paper of 1865

The universe may have begun with a bang, but it's going to end with a whimper. That's the point of entropy, at least as commonly understood. And indeed, early proponents of the concept, such as William Thomson (Lord Kelvin) and Hermann Ludwig Ferdinand von Helmholtz, warned that the universe is heading for a "heat death," when everything is the same temperature and nothing interesting will ever happen.

The concept traces to nineteenth-century advances in thermodynamics, the study of relationships between heat (thus *thermo-*) and work or motion (thus *dynamics*). The pioneers in the field—the Frenchman Nicolas Léonard Sadi Carnot, German Julius Robert von Mayer, and Englishman James Prescott Joule—had one common aim: building a better steam engine. Carnot, active in the 1820s, discovered that whenever heat is lost, it is possible to get work out of the process. Joule, twenty years later, discovered that the reverse is also true: Whenever there is work, there is also extra heat. Joule and von Mayer separately deduced what is now known as the "First Law of Thermodynamics": Energy can neither be created nor destroyed; it can only change form—say, from potential energy to work to heat and back again.

In 1850 another German, Rudolf Julius Emanuel Clausius

(long names were de rigueur), added a second law: Heat can never spontaneously pass from a colder body to a warmer one. This law may seem perfectly obvious—if you put water in an oven it will never turn to ice. But from the perspective of scientists trying to get work out of heat it had important consequences. The transfer of heat from one body to another is irreversible: The heat you use to cook a turkey can never be put back into the oven, at least not by the turkey. From Carnot's experiments, Clausius knew that work happens when energy (in the form of heat) passes from a state of greater excitement to a state of lesser—that is, from warmer to cooler. So heat can only be harnessed once to do work before it is spent—that is, dissipated into a cooler body from which it can't be retrieved without adding more energy to the system.

In fact, you'd have to add more energy to the system than there was in the heat to begin with. This is so because, as Lord Kelvin announced in 1851, there is always waste in any transfer of

Eureka!

heat. (Friction is a main, but not the sole, cause of this waste.) So much for the idea of a perpetual motion machine—to keep any machine running, you must keep pumping energy into it, because no mechanical process is 100 percent efficient. Every time energy is converted from one form (say heat) to another (say electricity), a little of it is lost to use. This principle is generally known as the "dissipation of energy": Even though energy can neither be created nor destroyed, it tends to dissipate, or pass from a more useful form to a less useful one.

To visualize why this is so, imagine an insulated room where the temperature at one end is 80° and at the other 20°. Obviously, heat will tend to flow from the hot side to the cooler until the whole room is one temperature (say 50°). What's happening is that the air molecules at the hot end, which are zooming around rapidly, start bouncing into the slower-moving molecules at the cold end, with the net result that zippy molecules give off energy to sluggish ones, and both bounce away at some intermediate speed. Given enough time, the speed of all the molecules will average out, meaning that the entire room will have the same temperature.

If you're smart, you'll find a way to harness the energy as it flows, via heat, from one end to the other. But once all the molecules are moving at virtually the same speed—once the temperature has evened out—you can forget it. This is because the motion of molecules will have become totally random. When one end of the room was hot and the other cold, there was a certain "order" to the state of affairs (most of the zippy molecules here, most of the sluggish ones there), and furthermore there was an order to how the energy was transferred (it moved in one direction). When the temperature is all the same, there's no longer any order (the remaining zippy and sluggish molecules are all mixed up), and no ordered transfer of energy in any direction. And to get the room back to its original state—say by heating one end and refrigerating the other—would take more energy than you could get back out of it if you did so.

Given the First Law of Thermodynamics, this all points to a depressing conclusion: The amount of useful energy in the universe is slowly but surely dissipating into useless energy. It's nice, for example, that the sun transfers its energy to the earth so that plants can grow and life go on, but sooner or later the sun will peter out. The world, the solar system, and finally the universe must ultimately all reach the same even temperature. When there's no difference in temperature (no organization of heat), no work or anything else requiring energy can happen.

Entropy is Carnot's word for the amount of useless energy in a system—the amount of "dissipated," disordered, or constant-temperature energy, which cannot be converted into work. And as he so bleakly put it, "the entropy of the universe tends toward a maximum."

But don't despair—heat death is a long way off, if indeed it ever happens. Kelvin and Carnot assumed that the universe is a closed system, but in fact the most recent theory portrays it as expanding and cooling. This isn't necessarily good news in itself, but it does complicate the heat-death picture.

What's more, the Austrian physicist Ludwig Boltzmann (1844–1906) showed that the Second Law of Thermodynamics is not the rigid, deterministic physical law Carnot thought it was. As he realized, the overall state of a gas (such as the 80° air in our room), measured by temperature and volume, is not uniquely determined by any particular pattern of molecular activity. That is, the molecules in our room could be zooming around in lots of different ways and we'd still get the same temperature.

Boltzmann called these scenarios—patterns of zooming molecules—"microstates," and defined a particular "macrostate" (temperature and volume) to be more "probable" the more microstates could produce it (the more likely it is to eventually appear in nature). Entropy, by this definition, measures the *probability* of a macrostate, and the Second Law thus asserts that systems tend toward a state of maximum probability.

Of course, this still sticks us with increasing disorder, if not with heat death, because disordered states are a lot more probable (easier to produce) than ordered ones. If you shuffle a deck of cards enough times the cards will be back in order, but doing so before doomsday isn't likely. Likewise, it's very difficult to unscramble an egg. But what Boltzmann's theory allows is a chance increase in useful energy without anyone's having to do anything at all—nature will occasionally *decrease* the entropy of a system.

And anyway, even if the universe as a whole is heading toward chaos, it's always possible to decrease entropy locally. Every time you clean up the mess in your office, you're producing order and decreasing entropy, and the same goes every time a plant blooms. Such order may come at the expense of disorder elsewhere, and of course useful energy is expended in the process. But as the sun isn't going to burn out anytime soon, you may as well cheer up and take what you can get..

Cybernetics

It is the thesis of this book that society can only be understood through a study of the messages and communication facilities which belong to it; and that in the future development of these messages and communication facilities, messages between man and machines, between machines and man, and between machine and machine, are destined to play an ever-increasing part.

Norbert Wiener, *The Human Use of Human Beings: Cybernetics and Society* (1950)

"Cyber" is a very trendy prefix these days, thanks to science-fiction novelist William Gibson, who coined the term "cyberspace" in the mid-1980s. Cyberspace is more than a new neighborhood in regular space; it's a quasi-reality, generated by telecommunicating computers, where people and data interact in strange new ways.

Given the prevalence of "cyber-" to name these new man/computer relationships, you may be surprised to discover that its root is the Greek for "steersman," *kybernētēs* (also the root of "governor"). And in fact its first use traces to the 1940s, not the 1980s, when an American mathematician at M.I.T., Norbert Wiener (1894–1964), coined "cybernetics" to describe the science of controlling or "steering" automated weaponry and other machines.

Wiener was particularly interested in the similarities between machine instructions and human language and behavior. He took as his starting point the use of "feedback" in automation. (His coinage of "cybernetics" was partly inspired by the first important essay on feedback, James Clerk Maxwell's "On Governors.") Feedback is, in a nutshell, what you get when you "feed" the results of a machine's actions back into the machine. For example, when sound produced through a microphone is channeled back through that microphone, we experience the joy of audio feedback.

But in most other cases feedback is desirable, as it enables machines to adapt rather than rely on fixed instructions. Central

Eureka!

heating, for example, works on the principle of feedback. Rather than rigidly pumping heat through your home at preset intervals, thermostat-controlled heaters are able to respond to changes in temperature, turning on and off as needed.

Feedback is one important element of Wiener's more general argument that machines and human beings handle messages in very similar ways. The central nervous system, for example, may be understood on the model of a feedback machine. The brain sends impulses, or "messages," to muscles in the hand telling it to move this way or that after receiving "feedback" from nerves in the hand. (The brain tells the hand to grasp a glass; the hand tells the brain that the glass is hot; the brain tells the hand to let go; and so on.) This continually looping process is the basis of both reflexes and learning.

The key similarity between man and machine, for Wiener, was that both (at least if the machines are "smart," that is, capable of channeling feedback) are able to organize things and generate information. He used "information" in a very broad sense, to mean something like "order." The second law of thermodynamics states that systems tend to become disorganized, static, and predictable—they are "entropic" [see p. 129]. If information is order, then it is the opposite of entropy; the more disorganized or predictable the message, the less informative it is. "Clichés," he points out by example, "are less illuminating than great poems."

What Wiener is getting at is that there is a unified science of information and control—cybernetics—that applies equally to human and machine communication. But for this science to have general rules, one must downplay the differences between man and machine. Notoriously, Wiener attempted to dispense with such concepts as "life," "soul," and "vitality" as purely semantic—"question-begging epithets," as he put it. For the purposes of the study of messages, he thought, it is best to avoid such concepts and "say merely in connection with machines that there is no reason they may not resemble human beings in representing pockets of decreasing entropy in a frame-work in which the large entropy tends to increase." It is even possible for machines, as feedback accumulates, to "learn"—one good example is the message-pad computer that gradually learns to recognize its owner's handwriting.

The cybernetic doctrine is the basis for more recent studies in artificial intelligence, or "AI." AI is possible to the extent that human thought can be modeled formally—that is, as a set of clear instructions for processing information. If the mind could be completely reduced to such instructions, then there's no reason a fully programmed machine couldn't be said to have a "mind." Many find the idea repellent, insisting that thought can't be reduced to mechanistic behavior. But nobody has so far been able to prove that mental phenomena—even such things as love and pain—are anything more than programmed nervous impulses. Some see a solution in Gödel's Incompleteness Theorem [p. 113], but not everyone is convinced.

The Big Bang

Most scientists are skeptical at best of the biblical myth of creation. But their favored explanation of where heaven and earth come from is, on the surface, also pretty implausible. In fact, when scientist Fred Hoyle coined its name, "big bang," he thought he was being funny.

The theory envisions a time some fifteen billion years ago when all the matter and energy in the universe—the entire "universe" itself—was concentrated in a single point of zero dimension and infinite density. In a sudden instant, before which concepts such as "before" and "instant" were meaningless, this point exploded out, unwinding itself within a fraction of a second into space and time as we know them. Over time, the universe continually expanded and cooled, allowing elements and later concrete objects to form. It has continued expanding and cooling ever since.

The logic behind this "big bang" theory is something like the logic behind the now discredited "Prime Mover" theory [see p. 5]. The story begins with Albert Einstein, whose general theory of relativity (1916) required that the universe be either expanding or contracting. At first Einstein was embarrassed by this corollary, since he, like all the astronomers of his day, assumed the size and shape of the universe were fixed or stable. So Einstein threw a cosmetic fix into the theory, a move he later regretted.

In fact, a mere year after Einstein published the general theory, the American astronomer Vesto Slipher published his own curious discovery that virtually every distant object he observed seemed to be moving away from the earth. The evidence lay in what is called a "redshift" in the spectra of light those objects emitted—their color as observed here on earth appears redder than the light those objects must emit.

The redshift can be explained by analogy to a passing siren. Everyone knows that the wail of an ambulance appears higher in pitch while the vehicle is approaching and lower in pitch as it recedes. (Pitch is determined by the frequency of a sound wave—that is, the number of times per second it achieves its greatest intensity.) This is called the "Doppler effect," which follows from the fact that sound-waves are "stretched out" if the source is receding. The same applies to light: If the source is moving away from the viewer, the light will appear to be of lower frequency, that is redder, than it would if the source were sitting still.

Slipher didn't quite know what to do with his discovery, but it made a lot more sense once, in 1929, astronomer Edwin Hubble announced a correlation between the *distance* of an object from earth and the *speed* at which it moves away from us. The two quantities, Hubble discovered, are in direct proportion: If object B is twice as far away as object A, then object B is receding at twice the speed of object A.

The logical conclusion to draw, if you consider relativity logical, is that the universe is constantly expanding. For a simple picture of why, imagine the surface of a balloon as someone blows it up. Both observation and geometry show that points on the balloon's surface move apart faster if they're farther apart. In fact, if point B was originally an inch from point A, and point C two inches from A, then C will appear to "recede" from A twice as fast as B does.

The same thing is happening with the universe, except it isn't a three-dimensional balloon, but rather a four-dimensional space-time continuum. This discovery of Einstein's was linked to Slipher's observations by a Belgian priest and math teacher named Georges Lemaître in 1927. And it was Lemaître who first tried to trace cosmic expansion backward. Just as Aristotle and his followers traced causes backward in time to an original "uncaused cause," Lemaître traced the expanding universe back to its origin.

Given that the universe is constantly expanding while total energy remains constant, the farther back we go in time, the more condensed the universe must have been. Both matter and energy—which by Einstein's theory are interchangeable—must have been concentrated in a smaller space. The earlier the state

of the universe, the denser and hotter it must have been, since heat measures the average energy contained in a given space. Carrying this process to its logical conclusion, we find all matter and energy concentrated in a single superheated point, which Lemaître called the "primeval atom." And he called the moment at which this atom began expanding the "big noise"—the name Hoyle later "improved."

It is of course difficult, if not impossible, to imagine such a beginning, for terms like "infinite density" and "singular point" defy everything we experience. Trying to imagine a time before time existed similarly dissolves into paradoxes. But perhaps it helps to think of Einstein's conclusion, in the general theory, that gravity is just a warping of the fabric of space-time. The denser an object, the more it "curves" the space around it, just the way heavier objects, when dropped on a taut rubber sheet, stretch the sheet more than lighter ones. The primitive, densely packed universe did not "contain" all space so much as it wrapped space around it into a point of infinite curvature. (The smaller a ball, the greater the curvature of its surface.)

None of this explains why there was a big bang; it just asserts that it must have happened. And what transpired in the seconds after the big bang is pure cosmological speculation, though evidence mounts every day (not all of it perfectly consistent with theory). Developed from the picture put forth by the Russian-American scientist George Gamow in the late 1940s, the standard description of the bang goes something like this:

At the moment of the big bang, there was only one kind of matter, called "superparticles," in the tiny universe. These particles collided violently for about the first 10^{-43} seconds after the bang—that's .00000000000000000000000000000000000000000000001 of a second. In this time, however, the universe expanded and therefore cooled to the point at which other particles could appear and be stable enough to resist the now less violent particle collisions. These new particles were the effectively massless electrons, photons, and quarks. By the time the universe was a second old, and still pretty hot—about 10,000,000,000 degrees Kelvin—a few other, bigger, more substantial particles were forming and surviving: neutrinos, protons, and neutrons.

Within about another 90 seconds, protons and neutrons began forming atomic nuclei, which developed into the earliest elements: first deuterium, then helium, then lithium and beryllium.

In their course, all the other known elements were formed, but this would take about a million years. What's missing from this picture, however, is the fate of those neutrinos (and the so-called "antineutrinos" born with them) generated in the first second of time. According to the theory, they should still be around somewhere in the background of the universe, though their radiation would have cooled by now to a very low temperature—about 2.7° Kelvin. This so-called "background radiation" predicted by the theory was in fact detected in 1965 by two Bell Telephone researchers, Arno Penzias and Robert Wilson.

This discovery was the first experimental confirmation of the big bang theory in general, and there have been many other confirming discoveries since that fill in some of the details. For example, in 1992 the American scientist George Smoot, by measuring radiation in Antarctica, showed that the "lumpiness" or irregularity of the universe was present in "seed" within half a million years of the big bang. It's difficult to understand, let alone explain, the import of Smoot's measurements, but suffice it to say that they allow us to explain the resulting shape of today's universe in terms of gravity, and they show that the big bang theory is definitely on the right track.

Chaos

Chaos, from the Greek for "gaping void," isn't necessarily a bad thing. As pure disorder there's little to recommend it; but what mathematician James Yorke meant when he borrowed the term in 1975 was patterned disorder—a shape underlying apparent randomness. And this is a very good thing.

"Chaos theory"—the study of such orderly disorder—became trendy only in the 1980s, but it originated in germ in 1960, as M.I.T. meteorologist Edward Lorenz developed computer models of weather patterns. As everyone knows, weather is very difficult to predict in the long run, even though we can isolate most of the factors that cause it. Lorenz, like others, thought all that was needed for better prediction was a more comprehensive model. So he wrote a program based on twelve simple equations that roughly modeled the main factors influencing the weather.

Lorenz discovered something surprising: Small changes or errors in a couple of variables yielded wildly disproportionate effects. Over the course of a couple of days, they hardly made a difference; but extrapolated out a month or beyond, the changes produced completely different patterns.

Lorenz called his discovery the "Butterfly Effect," taken from the title of a paper he published in 1979: "Predictability: Does the Flap of a Butterfly's Wings in Brazil Set Off a Tornado in Texas?" In other words, can minute factors eventually yield unpredictable, far-flung, catastrophic results? Lorenz indulged in a little hyperbole because he wanted to dramatize his point. Virtually all physics before the 1970s focused on what are called "linear" processes—processes in which small changes yield commensurately small results. But a great number of phenomena—not only in meteorology and physics, but also in biology, ecology, economics, and so forth—do not obey linear laws or follow linear formulas. "Nonlinear" processes are those whose equations involve variable rather than fixed rates of change, in which changes are multiplied rather than added, and in which small deviations can have vast effects.

The next step toward a theory of chaos came in the seventies when Yorke and his friend, biologist Robert May, began examining the properties of the so-called "logistic equation," which provides a simple model for population growth, among other things. (For the complicated details, *see* the EQUATIONS, p. 240.) The way this equation works is that results keep getting plugged back in to obtain new results, which are themselves plugged back in, etc. What's interesting is that depending on how you tweak a certain factor, the results either become increasingly predictable or increasingly chaotic.

But even the chaos of the logistic equation has its own sort of pattern. While you can't ever predict what a particular result of performing the equation will be, you know that it will fall in a particular range. (If you graphed the results, you would see a steady shape or pattern emerge.) Many other equations behave similarly, producing chaos with a shape—among these are equations modeling turbulence in liquids or the rise and fall of cotton prices.

Such equations are the flipside of Lorenz's weather formulae: Where cotton prices go on a particular day is unpredictable (or we'd all get rich playing the futures market); but the history of cotton prices shows a particular order. The name given to this order is "fractal": If you plot a diagram of price fluctuations from minute to minute, hour to hour, day to day, week to week, month to month, and year to year, the shape of the most coarse-grained diagram (the year-to-year map) will be reflected in the more finely grained diagrams (the month-to-month map on down). A fractal diagram can be blown up to any magnification you want, and it will strikingly resemble, and sometimes exactly reproduce, the shape of the larger picture.

Such behavior of the cotton-price curve was discovered in the early 1960s by the Lithuanian-born and French-educated American polymath Benoît Mandelbrot. While working for IBM, he discovered that other phenomena shared the fractal quality of cotton prices—for example, the distribution of "noise" (errors) in electronic transmission. Gradually, Mandelbrot found other examples of the same behavior, turning for example to geography for his groundbreaking paper "How Long Is the Coast of Britain?" The basic idea of this paper is that all sorts of natural objects, such as the British coast, have a degree of roughness that looks the same no matter how closely you  approach them. Seen from a distant point, or looked at through a microscope, a coast will look equally irregular—so that, lacking any telltale sign as to how far away a picture of the coast was taken, it would be difficult, if not impossible, to tell.

To describe this recursive, self-reflecting irregularity or roughness, Mandelbrot extended the notion of mathematical dimension. We're used to thinking in terms of integral dimensions—a line of dimension 1, a plane of dimension 2, a cube of dimension 3. But Mandelbrot introduced the concept of fractional dimensions—1.3, 2.7, 12.2—to describe the recurrence or roughness he saw in coastlines and price curves. (Think of a fractional dimension as a measure of how much of a total dimension a line or shape consumes. The rougher a shape, the more space it consumes.) In 1975, he coined the term "fractal" to name this new fractional-dimensional geometry.

Fractal geometry and chaos might have remained curiosities save for the discovery, by physicist Mitchell Feigenbaum in the mid-seventies, that a great many seemingly unrelated nonlinear systems behave in remarkably similar ways. This suggests that there might be a unifying theory to account for the chaotic behavior of systems and equations in a whole range of fields. And that's when scientists really started paying attention.

Chaos hasn't been around for long, and it's still being refined; new applications are discovered or invented, papers continue to be published, doubt and demonstration alternate rapidly. Still, chaos theory has shed light on the behavior of systems, quintessentially systems of flowing liquids, which are apt to change quickly from stable to seemingly chaotic behavior, the way water goes from steady to boiling as the temperature is slightly increased. (At 99.5° C, water is just hot water; at 100.5°, it is water changing state, becoming gas.) The jargon can be intimidating—such things as "strange attractors" are hard to explain. (They're basically shapes that constrain nonrepeating curves, if that's any help.) And such ideas as "fractional dimension" are apt to seem bizarre or uselessly abstract—but in fact fractal geometry has many practical applications. As author James Gleick points out in his popular book on chaos, measuring the fractal dimension of a metal's surface will tell you about its strength. The earth's surface has a fractal dimension, as do the blood vessels in your body. Even the human brain and its consciousness may have fractal shapes. Fractal geometry has been adopted in such places as General Electric, Exxon, and the Hollywood studios, not a group known for its indulgence of pure theory.

Do You Have Your Mother's Eyes?: Evolution and Genetics

Ontogeny Recapitulates Phylogeny

Ontogenesis or the development of the organic individual, as the series of changes in form which every individual passes during the whole period of its individual existence, is immediately conditioned by the phylogenesis or the development of the organic stock (Phylon) to which it belongs.... Ontogenesis is the short and rapid recapitulation of phylogenesis, caused by the physiological functions of inheritance (reproduction) and adaptation (nourishment).

Ernst Heinrich Haeckel, *General Morphology of Organisms* (1866)

Do species remain essentially the same in time or do they change? How does an organism grow from embryo to adult? These two distinct questions seemed to find mutual solution in the theory that "ontogeny recapitulates phylogeny" (a theory now out of favor).

In plain English, the idea is that the history of an organism's development (its "ontogeny") repeats the evolutionary development of its species ("phylogeny"). That is to say, if the evolutionary ancestors of human beings include fish and apes, then at different points in its growth a human embryo will resemble an adult fish and an adult ape. The idea was developed in the 1860s by German zoologist Ernst Haeckel (1834–1919), who called it the "biogenetic law"; the English summary, "ontogeny recapitulates phylogeny," dates to an 1872 issue of the *Quarterly Journal of Microscopical Science*. (Haeckel invented the terms "ontogeny" and "phylogeny," as well as the now more familiar "ecology.")

Behind Haeckel's theory was the ages-old question of how organisms take shape. As Aristotle, the first great zoologist, pointed out, animal embryos initially appear virtually formless. He was inclined to believe that growth occurs in three distinct stages, during each of which a new form is impressed from without on the embryo.

Against this theory, which predominated for centuries, another was advanced in the eighteenth century. Called "preformationism," it holds that organisms from conception contain their complete adult form, which is unfolded in time. Thus the human embryo from the very start has pairs of arms, legs, lungs, eyes, ears, and so forth, just in primitive versions. No shape or form need be imposed from without; everything is already there merely awaiting growth. This process, ironically, is what biologists originally meant by "evolution" (literally, "unrolling"), even though it contradicts what we now mean by "evolution."

Preformationism was falling out of favor by the turn of the nineteenth century, as philosophers, scientists, and poets alike began to view the world not as preformed or static, but as a dynamic process of constant, progressive change. At the same time, other "Romantic" ideas took hold, among them a belief in the essential unity of man with all of nature. Inspired by such ideas, a group of German biologists known as *Naturphilosophen* ("natural philosophers") first proposed a sort of biological recapitulation.

As they saw it, man was the greatest and most advanced of beings on earth, the goal toward which all Nature has ever striven and in whom it is united. Given that Nature operates, as they assumed, by universal uniform laws, man must represent the most advanced stage of an organic development shared by all creatures. All lower organisms, concluded the *Naturphilosophen,* were but partial approximations of man, and man the final stage in a process of perfection. Thus, as the human being grows from embryo to newborn, it must pass through all the lower approximations to reach the higher, while lesser animals are fixed in a state of arrested development.

This theory—which didn't exactly take the world by storm—still falls short of Haeckel's. For all the *Naturphilosophen* said was that the human embryo passes through the stages other organisms traverse *in the present*. What's more, while each "higher" species does represent a kind of evolutionary step beyond the "lower" species, species themselves do not change over time. Haeckel's more truly evolutionary version of the theory was inspired, of course, by Charles Darwin's *Origin of Species* (1859), translated into German in 1860.

On its face, there's nothing implausible about the thesis. Human embryos *do* retain features (such as gills) that are evolutionary relics and which are lost or superseded as the fetus develops. If nature is economical and doesn't come up with laws or processes it doesn't need, then it would make sense that the embryo's passage from simplicity to complexity would replicate the evolution of simple to more complex organisms. Haeckel in fact believed that evolution (phylogeny) directly *caused* the path of ontogeny.

But on further examination, the notion proved inadequate. The biggest problem is Haeckel's theory of evolution. In his view, a species evolves by adapting to its environment and then by passing on the resulting changes to the next generation. (This position is known as "Lamarckism.") When it was later shown that evolution depends on (essentially random) genetic mutation, often in the earliest stages of development, the rug was pulled out from under biogenesis. For if evolution "happens" (to speak loosely) toward the beginning of ontogenesis—that is, if genes mutate early in an embryo's development—then recapitulation fails. For Haeckel's theory basically states that phylogenesis is additive—that is, you take a series of evolutionary steps and add a new one at the end.

If anything, then, phylogeny recapitulates ontogeny. That is, it is when the development of an organism deviates from the normal path that evolution of the species becomes possible. These days, biologists look with more favor on the work of Karl Ernst von Baer, a German critic of the *Naturphilosophen* (and also, incidentally, later a critic of Darwin). In the 1820s, von Baer noted that embryonic development is not uniform or parallel among animals, but rather divergent. All vertebrate embryos, for example, look basically alike in the beginning, for they start out in their most generic, undifferentiated state. And if we compare the embryonic development of different species, we see that they don't follow parallel lines, but rather progressively deviate from the general to the particular, with the ultimate goal of producing an adult of that species. That is, each species follows its own, increasingly particularized path from egg to adult. *Divergence* from other species, not repetition of them, is the rule.

For these and other reasons recapitulation is officially rejected by biologists (though in rough outline it seems to have some truth). Nonetheless, Haeckel's theory spread through the sciences and humanities and has never been uprooted. Carl Jung, whose stock remains on the upswing, incorporated it into his theory of the "collective unconscious" [*see* p. 178]. No less an authority than Dr. Benjamin Spock stood by the idea in his popular manuals on child rearing, and no less respected a scientist than Stephen Jay Gould wrote a book in the seventies defending it. Recapitulation may be as hard to shake as the idea of progress itself.

Evolution ("Survival of the Fittest")

Though people argue heatedly over its validity, few question that the theory of evolution traces to the work of Englishman Charles Darwin (1809–1882). But while Darwin indeed gave it a firm scientific basis, he was hardly the first to propose it.

A century before Darwin the French naturalist Georges Buffon wrote extensively on the resemblance among various species of birds and quadrupeds. Noting such similarities and also the prevalence in nature of seemingly useless anatomical features (such as toes on a pig), Buffon voiced doubts that every single species had been uniquely formed by God on the fifth and sixth days of creation. Buffon suggested in guarded language at least a limited sort of evolution that would account for variances among similar species and for natural anomalies. But Buffon's suggestions were *too* guarded, and anyway the age was not yet ready to believe him.

A generation after Buffon, Darwin's grandfather, Erasmus Darwin, actually published an explicit theory of natural evolution, speculating in his book *Zoonomia* (1794–1796) that all living organisms had a common ancestor. The elder Darwin was on the right track, and he was able to draw examples from common experience (for example, selective breeding and protective coloration), but in the end his reasoning wasn't all that coherent. Indeed, it failed to convince even his grandson.

More germane were the theories of the French naturalist Jean Baptiste de Monet, the Chevalier de Lamarck (1744–1829). Lamarck did present a coherent theory, in *Philosophie zoologique* (1809)—namely, that species tend to adapt to the demands of their environment. So the giraffe, for example, developed a long neck because trees were tall in its native habitat; snakes lost their legs because they didn't need them to get around. In short, if a

Eureka!

living organism needs something to survive, it will develop it; and if it doesn't use a piece of its anatomy, it will lose it.

Lamarck had a great influence on the younger Darwin, but it was mostly negative. In fact, the theory Darwin eventually proposed, in *The Origin of Species* (1859), was diametrically opposed to Lamarckism. In Lamarck's scheme, when the environment changes, species evolve to survive; in Darwin's, species evolve as a matter of course, and the environment determines whether they survive or not. He believed in natural selection, better known as "survival of the fittest": Newly evolved traits will survive if they better "fit" a species to nature.

Darwin's theories originally grew out of a five-year charting expedition in the 1830s on the H.M.S. *Beagle*. Visiting remote corners of the globe, from the Cape Verde Islands to Brazil and New Zealand, collecting fossils and insects, studying geology, and taking copious notes, Darwin slowly formed his theory of evolutionary selection. Noting variations among the finches of the Galapagos Islands, he surmised that they all must have evolved from a single species, and that each new species was best adapted to one sort of diet. And much impressed with the evidence around him that the earth's surface had been gradually shaped over the course of centuries (by erosion, glacial passages, and so on), Darwin intuited that the earth's many species may have also reached their present state by a process of gradual evolution.

But this process, he thought, was far from placid. Influenced by the pessimistic ideas of Thomas Malthus, who painted human history as a competitive struggle for food and other resources, Darwin arrived at the idea that evolution likewise was a kind of competition. As new species gradually and naturally evolve, they find themselves competing with the old for food, territory, and protection against predators. Because natural resources are limited, and new species of life potentially infinite, nature must

impose a kind of limit on natural variation. Those best suited to nature's challenges and limits, Darwin concluded, survived to propagate their species.

But Darwin, a careful man diffident of his theory's acceptance, spent years breeding pigeons in the hope of producing plausible evidence for it. In the meantime, another young Englishman, Alfred Wallace, independently arrived at a theory virtually identical to Darwin's, and this spurred the latter to finally go public. Darwin presented a summary paper in 1858 and then hastily published his *Origin of Species* a year later; it was an instant best-seller. The evolution controversy had begun.

In addition to compelling theory, Darwin offered empirical evidence. He argued that atrophied organs, such as the human appendix and the penguin's wings, implied ancestral species that must have used them. He also noted that the embryos of verte-brates—mammals, lizards, and birds—are virtually indistinguish-able in the earliest stages of growth, and that the human embryo has a vestigial tail and gills.

Darwin heaped up plenty of other arguments, all of them cir-cumstantial but pretty convincing in sum. Yet copious evidence and careful argument hardly guaranteed Darwin a warm recep-tion. Few were publicly with him, and many against him, espe-cially as it seemed repugnant to human dignity (not to mention to religious faith) that man might have evolved from lower life-forms (the monkey was often cited).

But time, and further archaeological discoveries, would tell on the side of evolution. It still remains merely a "theory"—by its very nature, its secrets buried in time and its operations agoniz-ingly slow, it can never be "proved" the way mechanical princi-ples can. As long as belief in a literally true sacred scripture persists, evolution will never be accepted by all; and, indeed, there are still problems with the theory even on a scientific level. The theory has, like Freud's, had its ups and downs over the years, but it is likely that this great idea will, as it adapts to new data, survive in the competitive market of scientific ideas.

Mendel's Laws (Genetics)

Darwin's theory of natural selection was great so far as it went, but it quickly ran into a serious obstacle. According to Darwin's peers, traits are passed on from parents to offspring in equal measure: A smart mother and stupid father should thus produce children of average intelligence. This posed a problem for natural selection; for even if a "superior" individual appeared in a species, the superior trait would gradually be diluted through reproduction. Even Darwin was stymied by this, and in response he modified his theory, incorporating the Lamarckian proposition that nurture, as well as nature, must guide individual development.

Darwin, however, had supposed that evolutionary changes happen gradually; this hypothesis was soon proved false. Both William Bateson in England and Hugo de Vries in Holland discovered that species seem to evolve in sudden, discontinuous steps, called "mutations" by de Vries in 1900.

In the same year, de Vries happened upon some papers published a generation earlier by the Austrian monk Gregor Mendel (1822–1884). Though his work was ignored in his lifetime, Mendel, working with humble pea plants, had managed to uncover laws of inheritance that would revolutionize biology and lay the foundation for genetics.

Over seven years, from 1856 to 1863, Mendel crossbred and interbred plants with distinctive traits—tall plants with dwarfs, yellow peas with green, and so forth. He observed with surprise that such traits are not averaged or diluted, but remain distinctive: the hybrid offspring of a tall plant and a dwarf were always tall, not medium-sized. Yellow peas crossed with green yielded yellow, not greenish yellow, peas. Even more interesting, when Mendel interbred the tall hybrids, the subsequent generation retained the distinctive characteristics found in the "grandparent" plants: Most were tall, but about a quarter were dwarf. Likewise, third-generation plants of the yellow/green crossbreeding were 75 percent yellow and 25 percent green.

Mendel soon derived the math behind this phenomenon. Plants, like mammals, have two "parents," and each apparently contributes traits (tall or short, yellow or green) to subsequent generations. So while shortness may disappear in the second generation, it will reappear in some individuals of the third; thus the second (tall hybrid) generation must still carry "instructions" to produce short offspring. In fact, such instructions must come in pairs, one from each parent, and one element of the pair is passed on to each offspring of the third generation.

Mendel called this the "law of segregation": Inherited traits are passed on equally by each parent, and rather than mixing together they remain separate. That is, each trait is generated by a pair of instructions, with the "dominant" instructions determining how the offspring looks and the "recessive" instructions lying dormant. (Recessive traits appear only when both factors in a pair are recessive.)

Furthermore, according to Mendel's "law of independent assortment," which parent contributes which factor is governed only by the laws of chance—dominant factors are no more likely to be passed on than recessive ones. Inherited traits are also independent: The instructions for height have nothing to do with the instructions for color.

Though inheritance is usually much more complicated than crossbreeding peas, Mendel had stumbled on a fundamental genetic principle. Once Mendel's discoveries were married to cell biology, genetics emerged as a field. With improvements in microscopes, biologists were able to observe that cells reproduce by dividing in two, and that each resulting cell inherits half of each chromosome from the original. In the 1870s it was also discovered that when a sperm fertilizes an egg, the chromosomes combine.

These two observations together explained the basic mechanism of inheritance. Mendel's "factors" were eventually renamed "genes," and it was discovered that each pair of chromosomes in a cell carries several pieces of genetic information. By and large, genetics has hewed to a modified Darwinian line: Evolution proceeds sometimes by sudden (but small and gradual) mutation, with new characteristics being passed on genetically, but mostly by natural genetic variation (gene recombination). In either case, nature "selects" favorable changes for survival and rejects changes that are for the worse (as radical mutations generally are).

On the other hand, some biologists (for example, materialists in the former Soviet Union) took a more Lamarckian position: that environment (nurture) affects development and that changes due to environment are passed on genetically. Rigorous experiment has not corroborated the Lamarckian theory. Which in a way is too bad, because Darwinian evolution is rather harsh on species (such as the dinosaurs) that have not genetically adapted to a rapidly changing environment. The world would certainly be more interesting if, by a Lamarckian process, more species had been able to survive through the ages.

THE HUMAN SCIENCES

Oedipus, Less Complex: Psychology

A Pavlovian Response

The report does not hesitate to name names, a procedure that will inevitably touch off a Pavlovian response from Leftist circles to deride it as a "Reds under the Beds" scare.

The Daily Telegraph, Feb. 8, 1974

The name Pavlov may make you want to cry "Dogs!" We would call this a Pavlovian response, but that would be oversimplifying the point made by the phrase's namesake, the Russian physiologist Ivan Pavlov (1849–1936). Chiefly to blame is the British newspaper *The Daily Telegraph,* which in coining "Pavlovian response" in 1974 (phrases such as "Pavlovian conditioning" and "Pavlovian system" are older), used it merely as the equivalent of "predictable reaction."

Pavlov himself was actually more interested in unexpected or counterinstinctive behavior than in the predictable. He first made his mark (and won a Nobel Prize in 1904) with some glamourless but crucial work on the secretion of gastric juices. Pavlov discovered that while, predictably, the pancreas goes to work whenever you start chowing down a hamburger, it can also be set off just by thinking about a hamburger, or even by seeing a plastic model. He identified these curious latter instances as cases of "psychic secretion," laying the groundwork for his more famous later theories.

In a series of experiments that would appall contemporary animal-rights activists, Pavlov rigged up a few dogs to measure their secretion of saliva in response to various stimuli. Predictably enough (as per Pavlov's earlier research), the sight of raw burger set their mouths a-water. Pavlov then discovered that the dogs could also be made to salivate in response to any arbitrary stimulus—say, a sound or a kick—that they came to associate with the imminent introduction of dog chow. He called such arbitrary stimuli "conditioned" and the dogs' reaction a "conditioned reflex"—that is, a reflex artificially induced by training or habit. (This term first appeared in English in 1906 in the journal *Nature*.)

Not knowing when to leave well enough alone, Pavlov went on to extrapolate from these and other, more complicated observations a sort of grand psychological theory, attempting to explain almost all behavior, normal and deviant, in terms of acquired reflexes and their various interactions. After a brief vogue in the West, many of Pavlov's more grandiose claims were tossed overboard, but they were warmly embraced by the Soviets. Though not a Marxist himself, Pavlov's theories were virtually tailor-made to suit the Marxist view that human behavior arises out of the material conditions and patterns of life. If people are habituated to servitude by oppression, they can be molded anew once liberate. In other words, the Soviet system could condition a Soviet citizen. The results of this theory can be found in your newspaper.

Behaviorism

You walk into a movie theater and suddenly crave popcorn. You feel relaxed in a blue room and anxious in a red one. Feeling down in the dumps, you take a friend's advice and just try keeping a smile on your face; miraculously, you soon feel better.

How do we explain such things? Is there an objective way to speak about feelings? Do we need to refer to the "mind" or "unconscious impulses" to explain them? Or does it all boil down to a bunch of chemical reactions in the brain?

Behaviorism, generally speaking, is a school of psychology with particular answers to such questions. Unlike Freudians, this school has no use for hypothetical (that is, unobservable) concepts such as "the Unconscious" or the "id" in explaining psychic events. Taking what they consider a more scientific approach, behaviorists restrict themselves to observable data. And in the case of human psychology, what is observable is *behavior*—hence the name.

Behavioristic notions trace back at least as far as the writings of Thomas Hobbes, who viewed the human organism as a superior sort of machine. (In Hobbes's view, feelings and actions could be described as resulting from physical events or "motions" within the body.) But as a school and as a cause, behaviorism is essentially the creation of the American psychologist John B. Watson, whose 1914 tract *Behavior* announced its arrival.

Watson vehemently rejected the idea, held since Descartes, that mind and body operate according to different rules, and that the best (and really only) way to study the mind is through introspection. This, Watson claimed, was not science. First of all, introspection is *per se* subjective: Its findings cannot be objectively established. Second of all, introspection produces nothing even remotely like hard data: Its findings cannot be quantified. If psychology were to be scientific, said Watson, it would have to concern itself with hard, observable, and objective data. And it must leave aside vague (and he thought nonexistent) entities such as "consciousness" or "desire."

Very much along the lines of Pavlov, whose work with animals he only read later, Watson and his followers thought that scientific psychology lay in the study of relationships between external stimuli and individual responses. If we can show by experiment that some event (say, a bell ringing) regularly causes a particular behavior (say, a nervous twitch), then we've established a psychological claim. The total collection of such event/behavior associations suffices as a data pool, and only on such evidence are we justified in making psychological inferences.

Events become associated with behavior, the behaviorists say, through a process of learning or "conditioning." If a dog is regularly rewarded with a bone every time he obeys the command "Sit!" then he will learn that obedience is pleasurable and the command "Sit!" will thenceforth cause him to sit, almost as a reflex. (Behaviorist B. F. Skinner called this "positive reinforcement.") Similarly, if as children we learn that going to the movies means popcorn, we become conditioned to associate the event (going to the movies) with the behavior (eating popcorn), and the former will provoke an action to achieve the latter.

The basic idea of behaviorism, in short, is that behavior is not just a sign of some mental state but is in effect the *same* as a mental state. We don't get anywhere by concocting such absurdities as "temperament" or "id," which are just theoretical abstractions from how people behave. It is just as well, and more scientific, to ascribe such phenomena as "neurotic behavior" to conflicting reflex responses to overlapping stimuli. Besides, the behaviorist view supports the ultimate behaviorist goal: Their concern is not with theoretical models, but with making people act better. That is, if you can fix the environment, you can fix people.

Watson, of course, had to do some fancy dancing to explain those things most of us would view as much more mental than physical. In one absurd moment, he explained away thought as a kind of inaudible speech. (He first had to claim that speech is just conditioned behavior and not in any way "mental.") Emotions, too, were reduced to visceral actions.

Most implausibly of all, strict behaviorists had to cut meaning entirely out of their picture of behavior. They had no way of explaining how the same "stimulus" (say, the sound of a gunshot) might produce different "responses" in different contexts or at different times. By their model all that counts is the direct reaction to the sound, without reference to consciousness. But a gunshot means one thing on the street corner and another thing at the beginning of a race, and without some reference to consciousness of this difference one has no way of explaining why in one case we respond anxiously and in the other excitedly.

What's more, it isn't even clear to what degree animals, much less humans, learn through conditioning how to behave in the real world (as opposed to the lab). Without a basic stimulus/

response hypothesis, however, behaviorism has no real foundation. Recent attacks on the hypothesis have resulted in a dramatic decline in strict behaviorism. But in modified forms the idea still has some validity, and we must credit the behaviorists with establishing psychology as an authentic science, something Freud by himself could never have done. Simplistic as behaviorism might be at the macro level of mental events and emotional responses, its emphasis on physiology has led to advances at the micro level, in the study of brain chemistry and brain function. And sometimes smiling *does* help you feel better.

The Unconscious

We were accustomed to think that every latent idea was so because it is weak and that it grew conscious as soon as it became strong. We now have gained the conviction that there are some latent ideas which do not penetrate into consciousness, however strong they may have become. Therefore we may call the latent ideas of the first type *preconscious,* while we reserve the term *unconscious* (proper) for the latter type which we came to study in the neuroses.

Sigmund Freud, "A Note on the Unconscious in Psychoanalysis" (1912)

The essence of psychoanalysis, founded by Sigmund Freud (1856–1939), is the "Unconscious." (I follow the standard use of a capital *U* when naming the faculty.) A seething psychic mass of dangerous notions, this Unconscious causes all our problems. It is that part of the self our better self would rather forget. It is what gets revealed when we make a "Freudian slip"—it's the truth we wish hadn't come out, a truth we weren't even aware of. The Unconscious guides our dreams, acting out its forbidden wishes and plaguing us with terrors. The Unconscious just isn't nice.

Freud didn't invent the Unconscious; he just inflated it to its current proportions. The term was in fact freely bandied about in nineteenth-century psychology and philosophy, though some philosophers dismissed it as a chimera. The mind, they argued, is simply the same as consciousness; there can be no such thing as an "unconscious thought" since if it isn't conscious it isn't a thought and furthermore doesn't exist.

Freud, having worked with hypnosis, hysteria, and dreams, begged to differ. Say a patient is hypnotized, given an order, and then awakened. Some time later she carries out the order, not knowing why. How is this possible, Freud asked, unless the hypnotic suggestion exists somewhere in the mind, but outside consciousness?

This was nothing terribly new. Freud's real contribution was to mark off two different kinds of unconscious thoughts. The first are what most psychologists had been calling "unconscious": the sort—such as half-forgotten phone numbers and the names of people you meet at parties—that reside somewhere just below the level of consciousness, waiting to leap to mind. The second—such as childhood traumas and taboo wishes—lie deeper and are actively resisted by the conscious mind. Freud called the first sort of thought "preconscious" and reserved "unconscious" for the second.

Having been reared in the mechanistic tradition of nineteenth-century science, Freud imagined the mind as a kind of machine full of psychic energy. His early writings especially are studded

with such terms as "pressure," "energy," "dynamic," and other borrowings from physics, which make the mind look like nothing so much as a piece of hydraulics. In his early model, the "Unconscious" is the innermost and darkest mental realm, populated with energetically charged notions straining to reach the surface. The conscious mind, in turn, exerts energy to keep down or "repress" these unconscious thoughts. The mind is treated as a solid object in space with "regions" and "boundaries," a topography capable of illustration (Freud was fond of diagrams). The dynamical relations between the two major regions—the Unconscious and the "Preconscious–Conscious system"—generate psychical events. (Freud himself came to feel a bit embarrassed by his early materialism, but it has marked psychoanalysis ever since.)

For example, dreams are produced (almost like a chemical) when deep unconscious desires—the strongest of which date to infancy—mix with the preconscious "residue" of the day. But the Unconscious cannot fully emerge even in dreams, so effective is the mind's protective repression. As they mix with preconscious residues, or as they struggle up into conscious fantasies, unconscious thoughts must undergo a transformation in order to escape censorship. There are, Freud decided, two modes of transformation: "condensation" and "displacement."

Condensation, which amounts to a sort of unconscious logic, packages a constellation of repressed ideas or wishes into one cryptic theme (suitable for psychoanalytic decoding). Displacement, on the other hand, shunts the psychic "energy" bound up in unconscious thoughts onto safer, more admissible ideas somewhere down a chain of associations. Displacement, in Freud's view, accounts for much neurotic behavior—that is, behavior in which a lot of energy is directed toward what would seem a relatively unimportant thing.

In any case, Freud came to realize that his topographic model of the psyche had a fatal logical flaw. If the conscious part of the mind engages in repression, then its own act of repression should itself be conscious. But in fact we are unaware of repressing or resisting the Unconscious, or that we're condensing, displacing, and otherwise censoring it. In short, there is a part of consciousness that is *itself* unconscious: the repressive part. This part of consciousness is not itself repressed, or else it would be part of the Unconscious. That is, while everything repressed is unconscious, not everything unconscious is repressed.

Upon this conclusion, Freud knew he needed to develop a better map of the mind. The result is described in our entry on EGO, ID, AND SUPER-EGO, p. 171.

The Oedipus Complex

[When a boy] cannot any longer maintain the doubt that claims exception for his own parents from the ugly sexual behaviour of the rest of the world, he says to himself with cynical logic that the difference between his mother and a whore is after all not so very great, since at bottom they both do the same thing. What he has been told [about sex] has in fact revived the memory-traces of his early infantile impressions and desires, and thus re-activated certain feelings in his mind. In the light of this new knowledge he begins to desire the mother herself and to hate the father anew for standing in his way; he comes, as we say, under the sway of the Oedipus complex.

Sigmund Freud, "A Special Type of Object Choice Made by Men" (1910)

Freud first wrote in a letter of 1897 of the "riveting power of *Oedipus Rex*." So riveting is Sophocles' tragedy that Freud adopted its scenario to explain why we're all so neurotic.

In case you've forgotten the story, it goes something like this: Prince Oedipus of Corinth learns from an oracle that he is fated to murder his father and lie with his mother. Horrified, he flees the city for Thebes, but along the way gets into a brawl with passing travelers. One thing leads to another and soon Oedipus has killed them all.

As it turns out, one of those travelers was Oedipus' true father, King Laius of Thebes, who had abandoned him in infancy. So by attempting to escape his fate, Oedipus fulfills it, though he doesn't realize this until after being crowned the new king of Thebes and after marrying Queen Jocasta, who turns out to be his mother. When the facts come to light, a mortified Oedipus puts out his eyes.

This sorry tale, Freud continued in his letter, "seizes on a compulsion which everyone recognizes because he feels its existence within himself." What Freud felt, and ascribed to every male, was sexual desire for his mother and a murderous jealousy of his father. No matter that in the play Oedipus feels nothing of the sort and only gets into trouble by trying to avoid his fate. Freud grabbed the paradigm anyway and dubbed this unhappy compulsion the "Oedipus complex," a term that first appears in his 1910 paper, "A Special Type of Object Choice Made by Men." (It's always the *Oedipus* complex; *oedipal* is a stand-alone adjective.)

Freud's abuse of Greek legend succeeded the "Seduction Theory," his original theory of abuse. Noting that an alarming number of neurotic patients reported unwanted sexual experiences as children, Freud believed that psychological disorders could be traced to such "seductions." But when he looked harder at his patients' tales, he decided that some, if not most, of them were fantasizing rather than remembering childhood abuse—or rather, that they were recalling childhood fantasies.

This "discovery" (which has lately been widely attacked) led to a crisis in Freud's whole view of psychological development. Like virtually everyone else, Freud had believed that children have no concept whatsoever of sexuality, let alone any sexual inclinations, and that any exposure to adult sexuality is incomprehensible and frightening. But if, as in his new view, children are capable of sexual fantasies, then they are not really "innocent." Not only do they have sexual feelings, but these feelings are powerful and autonomous.

With this, Freud thought he had the answer to many common neurotic symptoms. Some of his patients desired only women who were already attached. Others sought lovers who would be likely to cheat on them. Such men seemed to be inviting enmity and jealousy. Another category insisted on seeing their love objects as unique and irreplaceable beings; yet another suffered fantasies of "saving" a woman from imminent loss of virtue. To Freud's mind, all these neurotic desires were merely extreme manifestations of the universal and "normal attitude in love." In the abnormal cases, his patients simply failed to overcome childhood emotions most of us manage to conquer.

These emotions are love of one's mother and jealousy of one's father. The mother is in fact "unique and irreplaceable"; the father is the actual or potential sexual competitor. When a boy, at the edge of puberty, first hears from his peers about sex, it is treated as something "dirty." Thinking of his parents this way is a shock, and as infantile sexual feelings for his mother come flooding back upon him, the boy develops fantasies of "saving" her from his father by taking his father's place.

What has been revived—and what is normally overcome when the teenager transfers sexual feelings to girls his own age— is the Oedipus complex Freud was looking for. In infants, the complex begins when the boy finds himself stimulated by his mother's stroking and caressing, causing his feelings for his mother to intensify. In time, his sense of excitement becomes centered on one very important part of his body: his penis.

Naturally, his parents aren't pleased when the boy starts paying undue attention to this part and one or both of them threaten to "take it away" if he doesn't leave it alone. The boy doesn't really believe this, until one day he spies the private parts of a little girl—and she's missing a penis! Her penis must have been taken away, he concludes, because she was having too much fun with it. The threat of castration all of a sudden becomes very real to him, and this shatters the Oedipus complex: If his sexual feelings and pleasure continue, his prized penis is doomed. His narcissistic attachment to it prevails over his urge to play with it.

As a result, Freud continues, the little boy learns to channel ("sublimate") his libido into less threatening feelings such as affection for his parents. Love of the mother and resentment of the father for "taking" her from him are overcome by identifying with both, which is how the little boy gets his "super-ego" [see p. 172]. In the meantime, his sexual desires and fantasies are repressed during what Freud calls the "latency period," until they are awakened at puberty.

You will notice that all this time we've been focusing on the little boy and his penis; what does Freud have to say about little girls? At first, he assumed that girls go through the same stages boys do, with the exception that their sexual fantasies are focused on the father. (Actually, Freud thought that infantile sexuality was to a degree bisexual, just that desire for the mother predominates in boys, and for the father in girls.) But after deciding that fear of castration is what kills the complex in boys, he had to change his model. Girls, after all, cannot fear castration—not because it's impossible but because they think it's already happened. (Freud has also been rightly slammed for supposing that girls see their genitals as "lacking," that they "envy" the boy's penis, and that they feel forever inferior for not having one.)

In his new, somewhat tentative and fuzzy model, Freud proposes that girls have a much simpler infancy. Rather than wishing to fulfill their sexual desires upon the mother as object, they wish only to take her place in relations with the father. Though, like boys, girls feel their first strong love for the mother, once they notice that little boys have something they don't, their love turns to resentment. Their father takes their mother's place in their affections, all the more so since once they accept the fact they'll never get a penis, girls begin to desire a substitute—namely, a baby. They wish to take their mother's place and give their father a child.

No fear of castration puts an end to this fantasy. Freud supposes that the Oedipus complex in girls is conquered by gentler means, namely "educative influences" and "external intimidation threatening the loss of love" ("The Passing of the Oedipus Complex," 1924). This, and the fact that girls' "pre-oedipal" attachment to the mother is extended and only superseded after an involved process, led Freud to reject the views of his protégé, Carl Jung. Jung held on to the belief that girls' development precisely mirrors boys', and in 1913 he coined the term "Electra complex" to describe the female version.

For this name Jung also turned to Greek tragedy, specifically to the several versions of the story of Electra. In Aeschylus's *The Libation Bearers,* Electra, prompted by the god Apollo, collaborates with her brother to do in their mother, Clytaemestra, who was responsible for their father's death. However, Jung, like Freud, seems to have relied more on Sophocles, in whose *Electra* the heroine takes the principal role in plotting vengeance. Though not as raw or suggestive as his depiction of Oedipus, Sophocles' story of Electra, as a woman deeply attached to her father and impelled to murder her mother, served Jung quite nicely as an analogue.

Ego, Id, and Super-Ego

In the earlier stages of developing a comprehensive view of the psyche, Freud proposed that the mind can be understood as a "system" with three distinct regions: the Conscious, the Preconscious, and the Unconscious. But by the early 1920s he had traded in this model for a new one.

Where Freud had once spoken of "the Conscious," he now spoke of the "ego," a term (from the Latin for "I") current in nineteenth-century psychology. The ego stands for more than the conscious self, though it includes it; in Freud's new model, the ego is that part or region of the mind that forms its outer surface and that develops out of our sense perceptions and experiences in the world. It is the ego that has conscious thoughts, but it is also the ego that (unbeknownst to our conscious minds) keeps threatening unconscious thoughts and impulses in check. The ego is the social self, the self most exposed to and influenced by "reality," in the form of sensations and social codes.

Obviously, if the ego develops out of experience, we are not born with it. The core of the psyche, which is later layered over by the ego, Freud calls the "id," from the Latin for "it." (He claims to have got the term from a contemporary physician, Georg Groddeck, who was influenced by Nietzsche.) This id is the dark, unconscious, libidinous center of interior experience, the locus of impulses and passions, the wild horse, to use Freud's metaphor, ridden and checked by the ego. While the ego busily (and unconsciously) represses the asocial and taboo impulses of the id, the id nonetheless manages to affect our behavior, channeling its energy into actions approved by the ego, or sometimes into actions we can't fully understand and that cause guilt.

Perhaps the easiest way to grasp the distinction and dynamic between ego and id is to pair up their contrasting characteristics.

The ego is coherent, the id incoherent; ego is rational, id irrational. Where the ego operates on what Freud calls the "reality principle" (responding to the demands and rules of the real world), the id operates on the "pleasure principle" (seeking to minimize pain and disturbance, which is to say to minimize psychic energy). The ego is at the surface of mental activity; the id lies at its depths. The ego is a mental representation of external sensation and experience; the id is a representation of instinct. The ego deals in concepts, principally verbal (what is conscious is that which can be spoken); the id deals in symbols, principally visual.

To the complicated give-and-take of ego and id, Freud added a third force, which is an outgrowth of the ego and which he calls the "ego-ideal" or "super-ego." The super-ego represents that which the self wishes to become, and it is the seat of such things as morality, duty, and faith. By Freud's somewhat overworked account, the super-ego arises just as the Oedipus complex dissolves. The way a little boy, for example, overcomes his illicit desire for his mother and hatred of his father is to "internalize" or "introject" these "objects" (mother and father). Love for the mother is then inverted into love of the self, or rather of that potential of the self to become ideal; and hatred of the father is vanquished by an intense identification with him as the higher or superior part of the self.

Thus the ideals, morals, prohibitions,

and laws of the parents take up residence in the mind, and forever after hold the ego to account. The super-ego is, in the end, the part of the self that is self-critical, that measures the distance between reality and the ideal, and that feeds off the energy of the id in order to channel the ego toward higher goals. Freud calls this process "sublimation," and to it he ascribes the achievements of human civilization and art, including presumably all the great ideas.

The Pleasure Principle

In the theory of psycho-analysis we have no hesitation in assuming that the course taken by mental events is automatically regulated by the pleasure principle. We believe, that is to say, that the course of those events is invariably set in motion by an unpleasurable tension, and that it takes a direction such that its final outcome coincides with a lowering of that tension—that is, with an avoidance of unpleasure or a production of pleasure.

Sigmund Freud, *Beyond the Pleasure Principle* (1920)

You may think of the "pleasure principle" as the impulse to wear party hats or polish off an entire pint of Ben and Jerry's. But Freud, who "discovered" it, thought we have the most fun when we feel absolutely nothing, especially not desire. Pleasure is a state where nothing ever happens.

This is because Freud thought of pleasure not as a positive feeling but rather as the absence of "unpleasure," or, in the charming German phrase, "*unlust.*" The psyche detests tension, which comes in many forms (anxiety, desire, guilt, etc.), and it instinctively wants to be rid of it. What we really want, and what the pleasure principle seeks, is a steady, undisturbed state, which Freud called "homeostasis." The pleasure principle is thus the psychological equivalent of the principle of inertia.

That we don't like tension might seem obvious, but Freud goes far beyond needing a back rub or craving nicotine. Just about any sort of disturbance, even one we *think* is pleasurable, is deep down very unwelcome. Any noticeable amount of "excitation," including sexual stimulation, is actually unpleasurable tension—lust is *unlust*. Even if we enjoy the buildup of excitement, it is only because we know its release will feel so good. And the more excitement or "tension" we release, the better it will feel.

Men will find this notion easier to accept than women will, at least when it comes to sexual excitement, but then Freud was notoriously clueless when it came to female psychology. (His

Eureka!

famous question "What does a woman want?" is justly held against him now.) In any case, while Freud always held to his position on the pleasure principle, his picture of it grew more complicated in time. Recognizing that while we have a strong tendency toward pleasure (stasis), our decisions or actions do not always lead us toward it, he concluded that other factors must be considered.

One of these is the "reality principle," the accommodation of reality in the service of self-preservation; often, pleasure must be postponed or sacrificed if we are to experience pleasure in the future. For example, while the pleasure principle might urge us to give the boss a good piece of our minds, the reality principle intervenes to hold our tongues—getting fired is a pretty high price to pay, on the "reality" scale, for easing a little tension.

Beyond the Pleasure Principle

Though the pleasure principle remained, to Freud's ever-changing mind, among the most basic of drives or tendencies, by the time of *Beyond the Pleasure Principle* he had concluded that there must be deeper impulses. One of these is the "repetition compulsion": a mysterious tendency to repeat or reexperience (in life or in dreams) even unpleasant experiences. One example would be the uncanny tendency of some people to always find themselves in abusive relationships.

Freud explains such strange behavior several different ways, all of them hypothetical. First, he describes the repetition compulsion as a return to the scene of an unsuccessful crime. The things we tend to repeat were originally very disturbing ("traumatic") experiences against which, at the time, we were unable to defend ourselves. We keep returning to them in an attempt to learn from our mistakes and retrospectively "master" the trauma. That way, the psyche hopes, we'll be ready for the next time.

Freud's second hypothesis is that repetition is simply inherent in life itself. As he said in his exposition of the pleasure principle, we like it when things stay the same and hate change or disturbance in general. Our instincts propel us toward the past, toward an earlier state of things we've been forced by external forces to abandon. Instincts are thus essentially conservative; they tend not toward change or development but rather toward sameness and repetition. There is "an organic compulsion to repeat," as in the way birds migrate every year and fish swim upstream to spawn in what may have been the ancestral home of their species.

Freud then takes this hypothesis to a new extreme. Not only do we wish to restore some peaceful past state, we ultimately wish to return to the most primal state of all—an inert or inanimate state, namely death. This "death drive," as Freud calls it, is to his mind the most fundamental of instincts. Even the instinct toward self-preservation, he proposes, is just an attempt to ensure that we die of natural causes, that is, die by an internal process.

Luckily for us, though, the death drive is only part of the ego. We have another deep instinct that counters it, namely the sex drive. Our sexual instincts aim not at death but at self-perpetuation and a kind of immortality. But the sex drive, a life instinct, is like the death drive conservative: It attempts to preserve life, but only by way of returning to a more primitive state (infancy)—that is, by "bring[ing] back earlier states of the living substance."

Even allowing for such life instincts, however, Freud can see no instinct in any life-form that aims at a new, different, or "higher" stage of development. Organisms may evolve, but only in response to external changes or pressures, not by any individual or collective will. At best, instincts may act to preserve obligatory (not willed) modifications in the structure or behavior of the individual or species. Above all, Freud sees no "instinct towards perfection at work in human beings," which he calls a "benevolent illusion"—a sentiment that has earned him few friends. Freud doesn't deny that some people strive tirelessly toward perfection, but he lays it all down to "sublimation," which is in fact the energetic repression of instinctive or unconscious drives.

Life and death instincts play themselves out at various levels of the psyche in a complicated dance. But in the end, as instincts, they share one goal in common: the restoration of an earlier state of things. This goal may or may not coincide, in some given instance, with the pleasure principle; but in the long run it seems that the pleasure principle, which seeks to resist stimuli, eliminate tension, restore equilibrium, and find peace, is closer akin to the death rather than the life instinct. The peace we deep down hope to find is the peace of the grave. Have a nice day.

The Collective Unconscious

We have to distinguish between a personal unconscious and an impersonal or transpersonal unconscious. We speak of the latter also as the collective unconscious, because it is detached from anything personal and is entirely universal, and because its contents can be found everywhere, which is naturally not the case with the personal contents.

Carl Gustav Jung, "On the Psychology of the Unconscious" (1943)

If you've read myths from different periods and cultures you've probably noticed some amazing similarities among them. So did the Swiss psychiatrist Carl Gustav Jung (1875–1961), who also noticed a number of basic, primitive images—as of demons, Earth Mothers, sages, and wild men—cropping up in sessions with his patients.

Naming these images "archetypes" (meaning "original impressions or patterns"), Jung concluded that they must be part of the unconscious mind that preexists any individual's personal experience. He called this part the "collective unconscious"—which does not refer, as you might think, to any sort of "group mind" into which we're all simultaneously hooked, but rather to a piece of each individual psyche that inherits the collective experiences and impression of ancestral humanity. (We all have an appendix, but there's no "collective appendix" we all share at once; we each have our own even while it is a vestigial, evolved, collective trait.)

Jung's theory of archetypes was just one element of a larger model that also involved the individual's basic "personality type" (extroverted or introverted), her "persona" (the self she shows to the world), her "shadow" (the self she represses), and her "personal unconscious" (the self she's forgotten about). But archetypes and the collective unconscious soon became, and remain, the most famous and most controversial elements of his theory.

Especially famous are two particular archetypal images—that of man-as-such (which Jung called *animus*) and that of woman-as-such (which he called *anima*). Having concluded that each person's unconscious is dominated by those traits and images excluded from the conscious self, Jung naturally assumed that the *anima* is the most powerful image in a man's unconscious and the *animus* the most powerful in a woman's.

Thus we are psychically hermaphroditic, and the harder we try to repress our hidden half the more likely it is to inflict us with psychological conflicts. In Jung's view, only by recognizing and accepting the contents of our personal and collective unconscious can we be psychically whole and healthy. This view led him to rely on "free association," which he believed would uncover for the patient those repressed, unconscious feelings and images that were so powerful and potentially threatening.

Though the psychiatric community was at first receptive to Jung's ideas, it came to feel they lacked a certain scientific concreteness and were too "literary." Jung in turn blamed science for alienating mankind from nature and its primitive forces by "explaining" everything in rational terms. In other words, it is science's fault that we're so out of touch with our ancestral experiences.

Lately, more psychiatrists have returned to Jung's writings, and archetypal psychology is enjoying a renaissance. (Ironically, there's as much of a literary sensation for archetypes too.) Clinical trials may never prove Jung's model correct, but as the men's movement and Joseph Campbell have shown, sometimes ideas don't have to be scientifically sound in order to satisfy the spirit.

Object Relations

The development of the infant is governed by the mechanisms of introjection and projection. From the beginning the ego introjects the "good" and "bad," for both of which the mother's breast is the prototype—for good objects when the child obtains it, for bad ones when it fails him.

Melanie Klein, "A Contribution to the Psychogenesis of Manic-Depressive States" (1935)

Freud thought he had figured most things out, but he never pretended to know what goes through an infant's mind. He did theorize about certain early psychic phases or "stages"—oral, anal, genital, and so forth—but thought it useless to inquire into such things directly. Freudian psychoanalysis only works when the patient is willing and able to hit the couch.

Among Freud's followers, however, the British analyst Melanie Klein thought it important and necessary to look to the very earliest of childhood experiences. The resulting theory is called "object relations," and it is central to contemporary psychoanalytic practice.

The theory takes its name from Klein's use of "object" to refer to the disjointed, fragmentary, and highly charged stuff of infantile perception. To the infant, the world is not made up of coherent people and things, which are distinct from the self or its needs and which may come and go, but rather of transitory objects that cause pleasure or pain. Objects that give pleasure are "good" objects; discomforting objects are "bad."

The quintessential "good object" in an infant's life is its mother's breast, the source of its greatest pleasure. In fact, the most compelling experience is feeding—taking in milk from the breast—followed closely by the discomfort of passing waste at the other end. These biological experiences provide us with our earliest psychological mechanisms, which Klein, following Freud, calls "introjection" (taking in or "consuming" external objects) and "projection" (sending or evacuating inner objects into the world). Naturally, the child wishes to "introject" or consume good (pleasurable) objects and to "project" or evacuate bad (painful) objects.

But nothing in life is that simple, so our earliest years are fraught with anxiety. Having no real sense of time or coherence, the infant doesn't know that pain (such as hunger or the burning of a rash) is temporary and will be relieved. Likewise, it is a shock when the mother takes her breast away and with it the heavenly state of pleasure. The slightest of changes can make a hell out of heaven or a heaven out of hell, which is all very trying and confusing to the little one's fledgling ego. The child discovers it has only very limited control over good and bad objects, and that it is difficult to take in and give out what it wants when it wants to.

A second complication arises out of instinct. Klein accepted Freud's theory that we are born with two basic and conflicting instincts: the death instinct (which is aggressive and destructive) and the life instinct (which is preservative and protective). A child's aggression is principally turned against those "bad objects" that cause it pain; what it doesn't know at first is that some of these bad objects are identical with good objects. For example, the hungry infant screams with rage when its mother's breast is dry; the breast is then a "bad object" that seems entirely different from the plentiful, pleasurable breast.

The infant in its misery wishes to destroy the bad breast and is consumed with violent fantasies. Eventually, though, it dawns on him that this bad breast he hates is just the same as the good one he loves, and that his aggressive, destructive impulses have been directed at the very source of his pleasure. Fearing that his fantasy actually threatens his objects and thus himself, the child endures an intensified inner war between destructive and preservative impulses. Out of this conflict emerges the famous "superego," that psychic agency bent on repressing dangerous instincts.

This picture of infantile fantasy, which Klein claimed to derive both from theory and from observation, reverses one of Freud's most troublesome hypotheses. According to Freud, the super-ego develops only after the Oedipus complex dissolves, say in about the child's fifth year. But Klein among others had observed a harsh and repressive psychic agency even in children of three, an agency even harsher and more repressive than the adult super-ego, not to mention the child's own parents. Object-relations theory not only explains the early appearance of a super-ego, it also explains its special harshness, which stems from the intensity of conflict among the infant's instincts and fantasies, which distort reality. Furthermore, Klein's theory does away with the embarrassing fixation of Freudian theory on the experiences of little boys. Both boys *and* girls feel pleasure and pain in equal measure, and they share, generally speaking, the same fantasies of consuming or destroying the "objects" of their disjointed world.

Thus, to close with Klein's own elegant formulation, "the formation of the super-ego begins at the same time as the child makes its earliest oral introjections of its objects" ("The Early Development of Conscience in the Child," 1933). And with this we leave the literature of psychoanalysis.

De-Signs of the Times: Postmodern Paradigms

Structuralism and Semiotics

A science that studies the life of signs within society is conceivable; it would be a part of social psychology and consequently of general psychology; I shall call it *semiology* (from Greek *sēmeion* "sign"). Semiology would show what constitutes signs, what laws govern them. Since the science does not yet exist, no one can say what it would be; but it has a right to existence, a place staked out in advance.

Ferdinand de Saussure, *Course in General Linguistics* (1916)

Structuralism (the study of cultural structures) and the related discipline of semiotics (the science of signs) were both given birth by a single text, the posthumous *Course in General Linguistics* of Swiss linguist Ferdinand de Saussure (1857–1913).

In practice, Saussure's two offspring are difficult to distinguish, but generally speaking, here's the scoop:

- Structuralism is the more inclusive term. It refers to the search for common "deep" structures underlying the broad range of cultural expressions. An example from anthropology is Claude Lévi-Strauss's analysis of myths: Lots of myths that seem different actually "mean" the same thing (serve the same function) since they operate the same way (have the same structure). Structuralists of one stripe or other may also be found among philosophers, historians, psychologists, and literary critics, in addition to linguists.

- Semiotics is a branch of structuralism. Its basic idea is that all sorts of behavior are communicative, which is to say that they "signify." Everything from the color of your tie to an act of war can be understood as a "sign" analogous to a word or sentence. Semiotics studies the systems, akin to language, in which such signs assume meaning. Most semioticians these days are either literary theorists or students of film, which accounts for their image as chain-smoking francophiles.

Saussure called himself neither a structuralist (the word hadn't been invented) nor a semiotician (though he coined that one), but in the *Course* he set forth the fundamental concepts common to each.

. He began by attacking the historical and comparative bias of contemporary linguistics. Linguists mostly busied themselves debating points on the history, development, and interrelationships of modern languages. (He called this historical/temporal aspect "diachronic.") Saussure contended that such studies put the cart before the horse, because linguists had no adequate theory of how language functions at any given time (in its structural or "synchronic" aspect). Imagine scientists studying human evolution without understanding biology or physiology, and you'll understand what Saussure was complaining about.

Saussure aimed to remedy this defect. First, he broke language down into two components: *langue* (the structure and rules of language) and *parole* (language as it is spoken). *Langue* is essentially synchronic—it is an abstract system fixed and invariant at any given time. *Parole* is fluid and diachronic—speech is the stringing together of words in time, and while governed by the conventions of *langue* it is ephemeral and prone to rapid change. (Think of *langue* as a huge, immaterial dictionary-*cum*-grammar manual.)

Saussure thought that to truly understand how language works we must first understand *langue,* which is more fundamental than *parole.* Here's his theory: *Langue* is a structure of signs that have no meaning inherently or in isolation, but only as part of the system. The word *tree,* for example, is a linguistic sign with a meaning for English speakers. But if you said "tree" to an Aleut Eskimo, it would have no meaning. A word means something

Eureka!

only if it has a particular place in a system of signs (for example, English). Furthermore, it doesn't mean what it means simply by being the collection of sounds that make up the word *tree,* but rather because it is *different* from all the other signs in English.

Saussure's ultimate point is that linguistic signs (sounds, words, phrases, sentences, etc.) have no essential meaning; in and of themselves, they are empty. Meaning is bestowed by the system of signs—the system of differences—which is entirely arbitrary. There's no necessary reason the word *tree* calls to mind the image of a tree; no reason the word "but" signifies contradiction; no reason adding the sound *s* to the ends of words in English makes them plural. All these meanings are defined by convention, as embodied in *langue;* meaning is a cultural product.

Which brings us to the point at which semiotics emerges from structuralism: the distinction between a *signifier* (a sound, mark, or cue) and a *signified* (the associated concept or image). For example, the sounds you make when you say the word *tree,* or the marks you make on the page when you write it, constitute a signifier; the mental image or concept of a tree is its associated signified. The combination of signifier and signified makes up the sign; and the arbitrariness of the sign follows from the fact that signifier and signified are only related by convention or common agreement.

Recognizing the arbitrariness of signs is only the first step of structuralist or semiotic analysis. The next step is to examine the system or structure behind them. The central figure of structuralism was not a linguist but an anthropologist, Claude Lévi-Strauss, a firm believer that you cannot understand any particular ritual, belief, practice, exchange, or myth unless you understand the total structure (set of patterns) of a given culture, which is hidden and unconscious. If you take a myth, for example, you can't understand it just by analyzing it in isolation—say by psychoanalyzing it or looking for its historical basis. What you need to do is look at all a culture's myths to discover the deep mythic "language" behind them all. This language is essentially bipolar—a set of oppositions (pure/impure, fertile/barren, raw/cooked, etc.) that get played out in various ways in every myth.

Semiotics, though foreshadowed in the works of Saussure and his contemporary, the philosopher Charles S. Peirce, really took off in the writings of the French critic Roland Barthes (1915–1980). Barthes also studied myth, but he extended the concept to include a wide variety of cultural codes and beliefs. In *Mythologies* (1957), he tackled the cultural significance of everything from Einstein's brain to professional wrestling, examining the ways objects and actions take on secondary or even

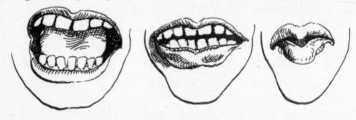

tertiary meanings in a culture. An example of my own: Certain colors and patterns on a piece of cloth signify a national flag; a national flag signifies national identity; national identity implies patriotism; patriotism implies obedience to the state; etc.

Barthes later trained his sights on literature, for example in his masterwork *S/Z* (1970), a study of Honoré de Balzac's story "Sarrasine." What Barthes wished to show is that what any given work "means" is partly determined by a wide variety of codes, some semantic, some ideological, some aesthetic, etc. Every text (or, if you wish, every author) attempts to place some limit on how such codes function, so that the reader experiences the feelings and meanings he or she was intended to. To the extent that a text succeeds in this project, it is "readerly"—geared toward passive consumption. But no author or work can rein in all the codes and limit the expansiveness or free play of meanings that exceed intention. To the extent that a reader participates in sorting out and assembling these superfluous meanings, he or she makes the text "writerly"—an object of active consumption.

During its heyday in the 1950s and 1960s, structuralism took root in a variety of humanistic disciplines, giving rise to schools of thought now called "post-structuralist." Movements such as philosophical deconstruction and Lacanian psychoanalysis have merited this dubious term because, while essentially structuralist, they question some of structuralism's assumptions—for example, the superiority of speech to writing or the coherence of the human subject. For further discussion of one of these fun and exciting movements, *see* DECONSTRUCTION, p. 191.

Universal Grammar

According to linguist Noam Chomsky (born 1928), perhaps best known today for his political lectures, the brain is no *tabula rasa* when it comes to language. The multitudes of human languages, living or dead, are just too similar in structure for coincidence. The brain, he thought, must come hard-wired with a "universal grammar" that enables children to learn language very quickly, but which also sets limits on what that language can look like.

Chomsky derived this idea from his study of syntax, which is the meaningful arrangement of words in a sentence. "The cat is on the mat" exhibits proper English syntax (and thus has a meaning), while "Cat mat the on is" does not. As Chomsky began his career, syntax was not a hot topic in linguistics; structuralism was in full swing, and structuralists cared lots more about the nature of the linguistic "sign" (word/concept pairing) than about grammar or the structure of coherent sentences.

Chomsky was also interested in structure, but not so much in the surface structure of language (the actual use of signs) as in what he called its "deep structure." Observing that virtually all children, whatever their native intelligence, easily and rapidly acquire basic competence in language, Chomsky theorized that human beings must share at birth some innate linguistic capacity. This is the capacity to learn, from hearing only a small number of all possible sentences, basic grammar and the rules for transforming sentences into new combinations.

Chomsky's point is that, since none of us learns the rules of grammar before learning how to speak, the brain must have a built-in grammatical capacity. When we teach a child the sentence "See Spot run," we don't (and don't need to) diagram the sentence into its grammatical components. The child somehow already knows that this combination of sounds is meaningful, and already has a sense of how the words fit together to make meaning.

What's more, a comparative study of the world's various languages shows that almost all of them favor a small group of common grammatical structures. The subject-verb-object combination, for example, is nearly universal. Even less simple structures, such as relative clauses, tend to look the same in every language. The Englishman's "The book that I read" is the Frenchman's "*Le livre que j'ai lit*": It's the same grammar. The Hebrew equivalent is slightly different—it would be "The book that I read it," a form we find in other languages (even at times in English). These two basic forms, "that I read" and "that I read it," describe the relative clause in practically every known language. Why only these two forms, when others could have done the same job? Why "The book that I read" and not "The book by me reading it past time done" or any other such formation?

Chomsky's answer is universal grammar: a grammar that both allows us to learn any language by example and limits the possible ways of forming a meaningful phrase or sentence. And though there are usually several ways of saying the same thing, each way must derive, by means of innate and fixed rules of transformation, from what Chomsky calls the "deep structure" of the sentence. For example, the spoken sentence "John is easy to please" is a transformation by strict rules from the more explicit and primitive sentence "It is easy to please John" ("move the object John to the beginning and remove the subject 'it'"). The similar-sounding sentence "John is eager to please" derives from a very different "deep structure"—namely, "John is eager to please someone"—by a different rule of transformation. That we instinctively grasp all these rules, and that we can understand many garbled or ambiguous sentences, supports the theory of a universal grammar.

If true, Chomsky's theory goes a long way toward explaining how human beings can ever say anything new. If language acquisition were purely empirical—in other words, if we learned all of language only from hearing it—then it would be difficult to explain how we could speak creatively rather than merely repeat what we heard. The capacity to substitute new words and ideas into learned sentence forms, at least, must be innate.

What's more, Chomsky believes, the deep structures beneath the surface of language must have some essential connection to the makeup of the brain. Thus, as he puts it, language is a "mirror of the mind" (*Reflections on Language,* 1975). In a very basic way, what we can think is connected to what we can say, not necessarily because thoughts and concepts are essentially linguistic (though some would say so), but because the brain is built to acquire speech, and the way it is built must determine the way we think.

Chomsky's ideas revolutionized linguistics and cognitive science, and they still play a large role in both fields. (But even Chomsky's views have evolved; he now advocates a version of his theory called "minimalism.") His theories do have their limits, particularly when it comes to understanding the act of speech, which interested him a lot less than the potential to speak. The dynamics of conversation and the subtleties of practical communication lie beyond any universal grammar; sometimes, surface is more important than depth. That we are all capable of speaking creatively doesn't mean we all do, and that we can say and understand "I love you" or "Beware the Jabberwock, my son" doesn't mean we will.

Deconstruction

Of course it is not a question of resorting to the same concept of writing and of simply inverting the dissymetry [speech over writing] that now has become problematical. It is a question, rather, of producing a new concept of writing. This concept can be called *gram* or *différance*. The play of differences supposes, in effect, syntheses and referrals which forbid at any moment, or in any sense, that a simple element be *present* in and of itself, referring only to itself. Whether in the order of spoken or written discourse, no element can function as a sign without referring to another element which itself is not simply present. This interweaving results in each "element"—phoneme [sound unit] or grapheme [written mark]—being constituted on the basis of the trace within it of the other elements of the chain or system. This interweaving, this textile, is the *text* produced only in the transformation of another text. Nothing, neither among the elements nor within the system, is anywhere ever simply present or absent. There are only, everywhere, differences and traces of traces.

Jacques Derrida, "Semiology and Grammatology" (1968)

Well, that clears everything up! But really, it's difficult to blame anyone for missing the point of deconstruction, since deconstructionist writing is so convoluted. It can get a little annoying, though, to see the word *deconstruction* popping up everywhere, with sneer implied, when most who use and abuse the term don't know what they're talking about.

Au contraire, some say; it's the deconstructionists, especially their leader, Jacques Derrida (born 1930), who are doing the abusing. Among deconstruction's sins, they say, are torturing language and denying all that is right and good. Deconstruction says there is no such thing as truth, that everything is relative, that morals are silly, and that meaning is up for grabs. Deconstruction, in short, is behind all evils of modern society, from political correctness to moral and aesthetic relativism.

There is a degree of truth in these distortions, but the panic is uncalled for. For the simple fact is that Derrida & Co. aren't really on a nihilistic campaign to destroy culture. They just want to destroy the metaphysical tradition in Western philosophy, which to their minds is a castle of nonsense built on air.

Here's the point in a nutshell, for those of you who can live without the details: Ever since before Socrates, philosophers have set up ideals such as "truth," "originality," "being," and "presence," and have proceeded to erect systems on these categories and their opposites. The philosophical dream has been to discover and clarify these basic metaphysical principles. But they're all full of holes, and any attempt to support them is inevitably rife with tautologies and self-contradictions.

Now for the dirty details; saner readers may wish to move on. A good place to begin is Derrida's best-known work, *Of Grammatology* (1967), which fires away at various of targets, including the structuralist writings of Saussure and Lévi-Strauss. Saussure, says Derrida, propagates the very old error of treating speech as purer and more original than writing. Speech is supposed to give presence and body to thought, while writing is a mere parasite, a lesser copy of speech. Derrida calls this doctrine *logocentrism*.

Saussure and his predecessors don't stop there. Logocentrism is merely a symptom of the metaphysical tendency to break reality down into paired opposites, one of which is "good" and the other "bad." Spoken language itself, in Saussure's account, is broken down into two opposed parts, called "signifieds" (meanings) and "signifiers" (sounds that point at meanings); signifieds are prior to and more substantial than signifiers, which are meaningless and arbitrary. A partial list of such oppositions includes:

speech : writing
signified : signifier
inside : outside
present : absent
presentation : representation
central : marginal

serious : rhetorical
being : nothingness
truth : falsehood
nature : culture

Metaphysicians view the world in such terms and consciously or unconsciously treat the first element in each pair as more original, purer, and better than the second. What they don't do is stop to question the logic of making such distinctions. Nor do they explain how such differences could have arisen in the first place if speech, nature, etc., were so pure, present, and good.

The aim of deconstruction is not to abolish such oppositions, which would be pretty quixotic. Nor is it to show that the second terms (writing, signifier, sensible, etc.) are in fact better than the first (speech, signified, intelligible, etc.), which would just be playing the same game backward. Rather, it is to show that the differences among the terms mask a mutual dependence or sameness.

Let's take the example of the speech/writing distinction. In the traditional philosophical account, speech is equated with the "presence" of a speaker, in several senses. The speaker is present in body as he speaks, but his speech is also the immediate presentation of his thoughts and feelings. Writing, on the other hand, is defined by the potential *absence* of the writer: You can read Rousseau's words even though Rousseau is dead. Rousseau's physical absence also makes discerning his intentions difficult: If one of his sentences seems ambiguous or confusing, you can't ring him up to ask what he meant. Deciding on the "true" meaning of his writing involves educated guesswork.

Derrida's rebuttal is that speech is no safer than writing from absence, confusion, or error. Speech is in fact all about absence, since if something were present to the eye or mind and its meaning or purpose were clear, then there wouldn't be any reason to talk about it. We speak in order to point at things—objects, ideas, attitudes, and so on—that aren't there or aren't clear. And since words don't summon the things spoken of into existence, they only reinforce their absence.

Though we speak in order to make present what is absent, to fill up space, and to cancel out silence and emptiness, in fact all we do is produce signifiers, not signifieds. This is what Derrida meant when he cryptically quipped that "*Il n'y a pas de hors-texte*" —"There is no outside-the-text." In his jargon, "text" means the tracing out of differences, the use of marks to separate *this* from *that* or to signify what is absent.

Speech is just as textual as writing is, as easily confused and as easily misinterpreted. We've all had those arguments that never end, in which our words create only more difference, not agreement. And speech, far from making things or ideas perfectly clear and present and thus shutting everybody up, seems inevitably to provoke more speech. How do I clarify what I mean? I use words, or signs. But are those signs clear? How do I clarify the signs? With more words, or signs. And so on and so on.

There is nothing outside this chain of absence and reference, according to Derrida. Even to speak of "inside" and "outside" is to invoke the same old categories that presume all the good things like presence and being have already been wrecked by bad things like absence and nothingness. That there was ever a time, ever a state of nature, free from absences, differences, and textuality strikes him as extremely dubious—a metaphysical fantasy.

Spoken or written, signs are self-replicating. Signifiers flow on like a river. There is no resting place in discourse, no last word, no transcendental anchor, no ultimate truth or presence we can summon to put a stop to all the talk and all the writing. How does one decide what is true, what is authentic, what is present except through discourse?

Speech and "writing" (in Derrida's broader definition) form an enclosed system whose only context is itself: If you look up a

word in the dictionary, you get more words, which you then look up only to get more words, and so on and so forth. Nothing outside language can guarantee its truth, authenticity, presence, or meaning; proofs, demonstrations, arguments, comparisons, and contrasts are all linguistic ("discursive") gestures. Derrida's point, then, isn't that nothing means anything, but rather that there's more "meaning" to go around than anyone can control.

Derrida "deconstructs" several other key metaphysical oppositions, such as center/margin and presence/absence. In each case, he shows that the first term is never self-sufficient, but always understood only in relation to the second term: A circle has a center only by virtue of having a "margin" or perimeter; a thing is present only in relation to its potential absence and the actual absence of other things. The ideas we consider "central" to the Western tradition are not so because they clearly and indubitably reveal a transcendental truth, but because people have made them so. They have become central by an *historical* process.

So Derrida does not deny that truth, beauty, and goodness exist—because they do exist, as functions, in all cultures. Rather, he denies that they are transcendental realities that exist outside and guarantee human concepts, discourse, and history. In short, people will keep talking and talking unto eternity, and no absolute truth or absolute presence will ever descend from above to stop them. Which is distressing/comforting all at once.

The "Global Village"

Postliterate man's electronic media contract the world to a village or tribe where everything happens to everyone at the same time: everyone knows about, and therefore participates in, everything that is happening the minute it happens. Television gives this quality of simultaneity to events in the global village.

Marshall McLuhan, *Explorations in Communication* (1960), Introduction

Marshall McLuhan (1911–1980) was so intent on explaining the whole of Western culture that he let a lot of details slip by. Even the generalities are a bit skewed and his reach beyond his grasp; but the main points are clear and the insights fascinating.

McLuhan prophesied a culture-shift unwittingly engineered by high technology, particularly electronic media. Its result is the "global village" he named in 1960. It is a world in which remote events can be communicated and experienced instantaneously, by radio, television, and, since his death, fax machines and computer networks. Space collapses, time is compressed, and with them dissolve the boundaries of such traditional and parochial worlds as neighborhood and nation. Our lives have become intertwined with those of all peoples.

This is an elementary observation, but McLuhan went much further. He expounds his larger theory in *The Gutenberg Galaxy* (1962), which begins with the introduction of printing to Europe. McLuhan claims that Gutenberg's invention did more than make books more widely available; it also revolutionized consciousness. The ancient Greeks, after adopting a phonetic alphabet, became capable of recording ideas in linear sequence, and therefore capable of linear, rational thought; printing spread it throughout Western culture. And by making texts readily available, printing fostered a new relation of self to society: Book cultures are individualistic and introspective cultures, in which the emphasis falls on "freedom of thought" and disinterested analysis.

Eureka!

In preliterate cultures the situation is entirely different. Culture before the book—oral culture—is centered on speech and hearing; knowledge is passed on by elders, and tradition (rather than individual initiative) dominates cultural activity. In McLuhan's analysis, experience centered on aural sensation is more concrete and more communal, more immediate and dramatic and emotional, than the primarily visual experience fostered by texts. A world of sound is a world of motion and activity, where all experience is concentrated in the present. The world of vision, on the other hand, is a cool world of distance and abstraction. Most of what we see, as opposed to most of what we hear, stays in one place and remains relatively unchanged. Most of what we see, as opposed to most of what we hear, is not addressed to us and does not directly concern us. We are at a greater distance from the visual; we can circle it, stand back from it, analyze it, put it in perspective.

Early in his career, McLuhan championed the rationalistic and linear values of book culture, which he saw as threatened by radio and television. In later works he confirmed the attack while giving up the value judgments. New technologies overwhelm us, they erode perspectives that depend on distance in time and space, they elude analysis and break open what seemed closed; but this isn't necessarily a *bad* thing. In a way, the fluid, disjointed, eternal present of the mass media experience, by erasing boundaries, brings us closer to an ultimate reality, which modern physics has shown to be constructed out of open fields and probabilities rather than fixed objects and certainties. In McLuhan's view there is something deeply impoverished about a culture fixated on the visual: Such a culture is poorly integrated, out of touch with the richness of experience communicated at once by *all* the senses.

Electronic media return us to the "village" in more ways than one. "Today," McLuhan prophesies in *The Gutenberg Galaxy*, "we move swiftly again into an auditory world of simultaneous events and over-all awareness." New technologies "retribalize" us, taking us back to a time when reality was more immediate and more fully textured, before the triumph of distance, abstraction, and linearity, so perfectly embodied in the mechanistic science of Descartes and Newton, and in the assembly line of the twentieth-century factory. The popular press, he says, "offers no single vision, no point of view, but a mosaic of the postures of the collective consciousness." Today, "our science and method strive not towards a point of view, the method not of closure and perspective but of the open 'field' and the suspended judgment. Such is now the only viable method under electric conditions of simultaneous information movement and total human interdependence."

One might hesitate to endorse McLuhan's enthusiasm. If all mankind is now one tribe thanks to CNN and the Internet, its clans are still capable of nasty behavior. Events prove that in the global village we're not yet one big happy family. It is perhaps equally alarming that the ritual magic of the tribe is now, by McLuhan's own admission, presided over by the priests of advertising. "Once more," he neutrally reports, "any Western child today grows up in this [tribal] kind of magic repetitive world as he hears advertisements on radio and TV." The global village is united not only by the instantaneous flow of information and images, but also by the ubiquity of McDonald's and *Terminator* movies. I think I'll go read a book now.

"The Medium Is the Message"

In a culture like ours, long accustomed to splitting and dividing things as a means of control, it is sometimes a bit of a shock to be reminded that, in operational and practical fact, the medium is the message. This is merely to say that the personal and social consequences of any medium—that is, of any extension of ourselves—result from the new scale that is introduced into our affairs by each extension of ourselves, or by any new technology.

Marshall McLuhan, *Understanding Media: The Extensions of Man* (1964)

If you find Marshall McLuhan's quip that "the medium is the message" a little confusing, don't be embarrassed—it *is* vague. McLuhan wants to make you puzzle out his meaning for yourself—that is, he wants to be "cool," a term I'll explain below. To translate simply, McLuhan is saying "the means of communicating (the medium) has more of an effect (message) than any information it communicates."

McLuhan viewed communication technologies as extensions of our bodies and senses—the camera is an extension of the eye, the radio an extension of the ear, and so forth. His big idea was that technologies are never neutral—that they aren't simply transparent vehicles for delivering messages. Rather, as they extend our senses they transform them, altering our relationship to space and time and affecting our interactions with the world.

The invention of writing, for example, not only extended our power to speak across space and time, it also made possible the development of rational and analytical thought, which transformed men's and women's relationships to nature and to each other. And the inventions of the printing press and the book, by fostering solitary reading and reflection, were essential to the development of individualism by the seventeenth century. The "message" (result, effect) of writing was analytical thought; the "message" of print was individualism.

McLuhan, though trained as an English professor, is most famous for his writings on electronic media, television in particular. He called television a "cool" medium, one characterized by intimate, low-definition images and ideas that require a high degree of audience participation. ("Hot" media like film and print present sharply defined images and encourage passive consumption.)

Though McLuhan agreed with critics who proclaimed that TV was radically altering society, he sneered at their moralistic attempts to censor or curtail certain types of programs. He claimed that the content of TV (the programming) is irrelevant; what is changing society, rather, is the medium's stimulation of new, more active ways of looking at the world, in which "information" is less important than patterns of feeling and engagement. His romantic way of putting this point was that television reintroduced young people to "mythic" thought, "the instant vision of a complex process that ordinarily extends over a long period of time."

Eureka!

Television, that is to say, negates space and time by bringing centuries of history and all corners of the globe into our living rooms every night. In other words, it has helped create a "global village." In the process, however, television "cools down" what it presents, taking the edge off conflicts and flattening personalities, making it possible to jump from war coverage to a beer commercial. The low-definition, participatory, process-oriented "message" of TV is perhaps best exemplified by such successful programs as "The People's Court" and "America's Funniest Home Videos"; its logical conclusions are those elusive "new wave" ads and that glowing parade of mythic patterns, MTV.

McLuhan prophesied that television's transformation of society would ultimately render print obsolete—and his prophecy may be coming true with the advent of computer books and the slow death of the newspaper. (Those still alive look more like TV screens every year, with their grainy color photos and infotaining graphics. The *USA Today* vending boxes are in fact modeled on TV sets.) But he also said that when the television image achieved the sharpness and definition of film, it would no longer be television, because it would no longer be cool. Had he lived to see HDTV (he died in 1980), he might now be predicting the death of the tube as we know it.

Virtual Reality

"Virtual reality" is a phrase on virtually everyone's lips these days, but does it really exist? No—at least not according to any dictionary I've seen. As language, then, it's just virtual, and the same might be said for its reality. The closest most people can yet come to virtual reality is a glorified video game in some self-styled "virtual arcade."

So what's the point? The point is that it's coming soon to your living room. As computers become faster, more powerful, and cheaper by the day, they open up more and more possibilities for ever-more-convincing simulations of real-life situations. The computer-generated reality of the future won't be just a high-definition picture. It will be, the prophets say, a high-definition picture that you can virtually walk into.

The term "virtual reality" (or "VR") developed out of computer jargon, in which "virtual" originally meant "nonexistent, but made to appear real by software." (It was first used, in 1959, in the phrase "virtual memory," which is still current and which means "hard-disk space appropriated by the system software for use as if it were random access memory.") Thus virtual reality, a sensory experience created by computer software—a three-dimensional simulation that, at its best, looks, feels, sounds, and smells like reality.

Like many new technologies, virtual reality or "VR" developed in the American defense industry. The first approximations were flight simulators developed to train combat pilots in the late 1940s. Using these simulators, pilots could master various dangerous scenarious without risking their actual persons (or expensive aircraft). A decade later, according to Howard Rheingold in *Virtual Reality* (1991), Hollywood cinematographer Morton Helig conceived and first built his "Sensorama Simulator," an arcade booth outfitted with handlebars and a faceplate for the virtual experience of (among other things) riding a motorcycle through Brooklyn.

But Helig's contraption, which went beyond vision to encompass all the senses, never got off the ground. It was only in the 1980s, when the Defense Department and NASA recognized the potential of VR (which had been brought up to speed at M.I.T. and elsewhere) that the technology drew serious attention and money. What takes VR beyond simulation is that rather than merely imitating or simulating an experience, it simulates the environment and conditions that make possible creating actual new experiences. It's a reality that happens in a virtual environment.

Every day seems to bring the reality closer. As yet the basic technology of virtual reality—helmets, "power gloves," goggles, and other computer-wired paraphernalia—is relatively crude, especially in its commercial applications. But there are definitely big ideas for future applications: virtual buildings you can "walk" through before they're built; virtual long-distance communication, in which you can reach out and touch someone thousands of miles away; virtual manipulation of molecules with one's power-gloved hands; virtual travel through a patient's body; virtual business conferences. (Kiss those junkets to Vegas and Frisco good-bye.) The big idea on most people's minds, of course, is virtual sex; but I wouldn't count on computer simulation being that good (alas) in our lifetime.

In Search of the Invisible Hand: Economics

Gresham's Law

Gresham's law—"Bad money drives out good"—sounds simple, but what does it mean? Usage varies, but the main sense today is that wherever people with uplifting intentions invest, sooner or later bad or ignorant people will take over. Smart money gives way to fool's gold.

But that's hardly what the "law" originally meant, when Sir Thomas Gresham (ca. 1519–1579) was England's top businessman. Gresham was celebrated in life and in legend as an entrepreneurial hero, on top of being an important government bureaucrat. Among his services to the Crown, he was in the early 1550s the royal agent in Antwerp, Europe's principle trading center. As his duties included handling the Crown's debt and negotiating currency exchanges, Gresham quickly became an expert on exchange rates and on the circulation of currency in general.

As legend has it, Gresham impressed himself on Queen Elizabeth I almost as soon as she took the throne in 1558. He wrote her that England's currency (and thus its goods) was suffering on the foreign market. Gresham tactfully blamed the situation on her predecessor "Bloody Mary," who had sanctioned the debasement of coinage.

Since nobody wanted the debased coins, which were made of cheaper metal than those already in circulation, everyone attempted to pass on the "bad" money while hoarding the "good." Thus the better coins were put out of circulation and the worse coins were being used to set the commercial exchange value. Put simply, as Gresham said, "Bad money drives out good." (Actually, Gresham was only quoting a proverb, but that didn't stop economist Henry Macleod from first calling it "Gresham's Law" in 1858.)

Gresham would go on in 1568 to found London's Royal Exchange, a place for merchants to meet, rub shoulders, truck, and barter. (He modeled it on Antwerp's Bourse.) The building was adorned with numerous stone grasshoppers, an insect central to Gresham's crest, and this gave rise to the tradition of English bankers, goldsmiths, jewelers, and other traders in money and specie placing grasshoppers on their signs.

Gresham would also be celebrated in popular legend for one spectacular display of his age's taste for conspicuous consumption. It is reported that when Queen Elizabeth paid her first visit to Gresham's new Exchange, the financier raised a cup of wine in which he had pulverized a stone worth £15,000. Even stranger, he drank the stuff.

Laissez Faire and
The Law of Diminishing Returns

Literally "let be" or "let do," "laissez faire" was an economic doctrine the French cooked up in the mid-1700s. At the time it referred to a noninterventionist policy, the economic equivalent of Henry David Thoreau's aphorism, "That government is best which governs least." Today, still left untranslated, the phrase is applied more generally to any hands-off attitude.

The original doctrine was advanced by a group of French economists subsequently dubbed "physiocrats" after the Greek for "the rule of nature." Led by François Quesnay (1694–1774), they believed that Nature is smart and good and knows what it's doing, while governments are fallible, easily misled, and often stupid. Yet the prevailing practice in Europe at the time was for governments to micromanage virtually every aspect of the production and distribution of goods. Government, the physiocrats argued, should "laissez faire" so that Nature might take its course, guiding the economy efficiently to its happy natural state.

Who actually coined the phrase "laissez faire" is a matter of debate. Some credit Quesnay, others Vincent de Gournay, a product inspector in the government of Louis XVI and a convert to physiocratic doctrine. Whatever the truth, the physiocrats were better at coining phrases than at instituting their policies. There was a brief vogue in Louis's court for laissez faire, but that didn't last long, and it was only reintroduced (in a somewhat crippled form) after the French Revolution. The doctrine proved more successful in England, where it took a central place in Adam Smith's groundbreaking *Wealth of Nations* (1776), the first text of "classical" economics.

The Law of Diminishing Returns

Another physiocrat, Robert-Jacques Turgot, is credited with formulating the "law of diminishing returns": After a certain point, continued effort or expenditure will produce lesser results.

Say, for example, that you discover a vein of gold running under your garage. It will profit you nothing unless you sink a chunk of money at first into leveling the garage and setting up for a dig. Thereafter, your profits for every dollar spent will be enormous, until, as the vein is exhausted, each dollar spent begins to yield less and less gold. Your mine has become victim to the law of diminishing returns.

Turgot formulated the notion, circa 1767, in terms of piling weight on a tight spring. Some significant weight is required to overcome the spring's initial resistance, but beyond that point even small pressure will compress it to some degree. However, Turgot said, "after yielding a certain amount it will again begin to resist the extra force put upon it, and weights that formerly would have caused a depression of an inch or more will now scarcely move it by a hair's breadth. And so the effects of additional weights will gradually diminish" (*Observations sur un Mémoire de M. de Saint-Péravy*).

The "Invisible Hand"

Every individual necessarily labours to render the annual revenue of the society as great as he can. He generally, indeed, neither intends to promote the public interest, nor knows how much he is promoting it. By preferring the support of domestic to that of foreign industry, he intends only his own security; and by directing that industry in such a manner as its produce may be of the greatest value, he intends only his own gain, and he is in this, as in many other cases, led by an invisible hand to promote an end which was no part of his intention. Nor is it always the worse for the society that it was no part of it. By pursuing his own interest he frequently promotes that of the society more effectively than when he really intends to promote it.

Adam Smith, *The Wealth of Nations,* Book IV, chapter 2

In 1776, the year of the biggest tax protest in history, Scotsman Adam Smith (1723–1790) published the work that popularized and propagated "laissez faire" doctrine. Fittingly enough, Smith has become, in the two centuries since, the patron saint of conservative free traders who hate no word more than "taxes."

Smith's work, *An Inquiry into the Nature and Causes of the Wealth of Nations,* is probably *the* most famous economic treatise, more for its clarity and rhetorical power than for its originality. Drawing on the ideas of the French physiocrats and of English forebears such as Sir William Petty and Sir Dudley North, Smith marshaled a series of arguments that dealt a virtual death blow to managed economy.

Smith believed, like many contemporary philosophers, that nature is man's best guide. God ("Providence") has so arranged things that if men and women are free to pursue their own proper interests, they will naturally act for the best of society. Whether they intend to or not—and mostly they don't—people help each other by helping themselves; even the greediest of motives often lead to the happiest results for all. This is the work of Providence's "invisible hand."

Smith first introduced the concept in his *Theory of Moral Sentiments* (1759) but only fully developed it in *The Wealth of Nations,* where he proposes the harmony of self-interests. If everyone looks out for himself, the result won't be Hobbes's "state of war," but rather a happy tide that lifts all boats. By enriching themselves, people enrich society, so society should allow people to enrich themselves as greatly as possible. Each person will do what he does best to realize the greatest profit, producing goods that others find cheaper to buy than to make themselves. And if anyone gets too greedy, benefiting at society's expense by driving up prices, the invisible hand will prompt others to join the fray and compete. That way, prices are controlled and unprofitable businesses crushed.

Since the invisible hand does its job so well, Smith concluded, it is folly for governments to intrude on production and trade. Furthermore, "kings and ministers...are themselves always, and without any exception, the greatest spendthrifts in the society." In short, "laissez faire."

This doctrine works well in a relatively free, strong, and expanding economy, such as that of Smith's England. But as experience shows, laissez faire doesn't work so well in a contracting economy, or in developing nations, or when unemployment is high. I doubt Smith would be able to explain why the invisible hand spins the business cycle downward or why it punishes free-trade booms with inevitable busts.

The Division of Labor

The greatest improvement in the productive powers of labour, and the greater part of the skill, dexterity, and judgment with which it is any where directed, or applied, seem to have been the effects of the division of labour.

Adam Smith, *The Wealth of Nations,* Book I, chapter I

Nobody invented or discovered the division of labor, which is simply the parceling out of pieces of work to different persons or groups. Dividing up complex tasks traces to the dawn of society and was effected on a grand scale at least as early as the construction of pyramids in Egypt. But before the Industrial Revolution it was not the standard operating procedure in the workplace, and we can thank nineteenth-century improvements in technique for modern mass production.

An important contributor to the eventual dominance of division of labor was Adam Smith. Smith considered it important enough to begin his *Wealth of Nations* on the topic. As an example, he cites the making of pins, a seemingly simple task, yet a slow and tedious one for a single person to perform. But if the manufacture is divided among ten to twenty workers, each highly skilled at one operation in the process, then it is possible to churn out about twelve pounds—48,000 pins—per day. Smith finds this quite exciting.

Smith's first point is that the division of labor allows each worker to focus on one task and one task only, at which he will naturally become highly dexterous. Second, time is saved if workers do not have to switch tasks. Last—and in retrospect most important—if labor is divided into a series of limited and precise tasks, it is possible to at least partially mechanize them, "enabl[ing] one man to do the work of many."

Of course, division of labor also has its costs, some of which Smith recognized. Since unemployment wasn't a real problem in his day, he could not foresee a time when mechanization might cost jobs on a massive scale. But he did worry that forcing laborers to spend each day in simple repetitive tasks might not uplift the spirit. If a man is not called upon in his employment to "exert his understanding, or to exercise his invention" in facing challenges, then he "generally becomes as stupid and ignorant as it is possible for a human creature to become." Smith's answer: free, and perhaps compulsory, education. Graduates of public schools in America today can best attest to the efficacy of this solution.

The Paradox of Value

No doubt you've wondered, as some trinket caught your eye, "How could *this* piece of junk cost $74.95?" Adam Smith wondered in much the same way why diamonds should be so dear and water so cheap when the former are nearly useless and the latter so essential. Of course one is rare and the other plentiful, but Smith realized that supply and demand would only account for *price* and not for *value*—that is, for why there would be a demand in the first place.

This is Smith's "paradox of value"; his answer is that diamonds are valuable because they take so much labor to extract, cut, and polish, while water can be had with a pail. In other words, value is produced by human labor; this is why a hand-carved cabinet is more "valuable" than one that rolls off an assembly line.

Of course, in some sense water is still more "valuable" than diamonds, because it is more useful—it has a greater "use value" though a lesser "exchange value." What Smith established with his labor theory of value was the difference between the two, with the latter serving, in conjunction with market demand, to determine a commodity's price.

Smith's theory led him to invest his faith in a free and open market as the best of all economic arrangements, since it forced producers, through competition, to minimize labor and thus cost. Ironically enough, Smith's labor theory of value was also the basis for Marxist economics. In Marx's view, while human labor does indeed create value, greedy capitalists will, if let be, inevitably pay their workers less than their labor is worth, pocketing the change. The difference between the cost of producing a product and its price on the market is called "surplus value." The moral of all this is that the way to solve Smith's paradox of value is to invent as many different kinds of value as necessary.

Dialectical Materialism and the Class Struggle

As you clever readers may have guessed, dialectical materialism is what you get when you cross materialism with Hegel's dialectic [see p. 51]. We owe the wedding to Karl Marx (1818–1883), who fell deeply under Hegel's influence as a young man but grew out of Hegelian idealism in a hurry.

A lot of people have the wrong idea about Marx. As we shall see, he didn't think capitalism is something a nation can do without. The materialist dialectic is about the gradual and regular march from oppression to liberty, from feudalism to communism, and capitalism is a neccesary stop along this path.

Now, by "materialism" Marx didn't mean a lust for possessions. He meant that human attitudes, aspirations, and activities are shaped by material circumstances (such as geography and economy). As for the "dialectical" part, Hegel had pictured history as an ongoing and progressive struggle, in which theses clash with antitheses to produce better syntheses. Marx liked the model, but he rejected Hegel's assumption that the historical dialectic is guided by "Ideas" or "Spirit"—i.e., God.

History does indeed progress through a series of reversals and upheavals, Marx believed, but these are inspired by the material circumstances of life, whose "base" is the economic structure of the time. Economic arrangements determine every form of cultural expression and change, from politics and class to art and religion. Marx called these expressions "superstructures."

Dialectic arises out of conflicts inherent in economic systems. In the capitalist system, to take the best known example, there is inevitably and necessarily conflict between those who control the means of production (the bourgeoisie) and those who actually produce (the proletariat). But capitalism is a necessary stage of economic development. Sprung from the ruins of feudalism (which met its antithesis in the development of the bourgeois class), the capitalist system promoted the development of industrialism and efficiency in production. But the internal conflict between capitalist and worker will lead inevitably to class struggle, with the workers coming out on top. In dialectical terms, the "thesis" of capitalism meets its "antithesis" in an organized proletariat, and from the resulting struggle emerges the "synthesis" of the socialist system.

Marx fancied that Europe—wracked by revolutions in the earlier 1800s—was right on the verge of achieving this synthesis. (As he saw it, such backward societies as Russia and China had a long way to go, since they hadn't yet gone through the capitalist stage.) He was less than prescient on this score, but on the other hand, he was under no illusions as to the form socialism would initially take. He foresaw a "dictatorship of the proletariat," which he thought inevitable but by no means ideal. In time, when private property had been completely abolished, class distinctions had disappeared, and thus society had become inherently just (according to materialist principles), then a new synthesis would emerge: that of the classless democratic society.

In other words, Marx would have viewed most of the political systems we call "Marxist" as a temporary, if necessary, evil. Decades of oppression and corruption were not any part of his plan. In practice social philosophies do tend to go awry, but as a method of historical analysis Marxism still survives in various forms. Anthropologists, historians, political scientists, and literary scholars still use his analytical tools to peel away layers of sentiment and idealism from the cultures they study. In fact, almost any analytic method emphasizing material conditions might be called "Marxist," which means that Marxism comes in many forms, not all of them compatible. So if the dialectic survives anywhere, it is within Marxism itself.

"Religion Is the Opium of the People"

Religious suffering is at one and the same time the *expression* of real suffering and a protest against real suffering. Religion is the sigh of the oppressed creature, the heart of a heartless world and the soul of soulless conditions. It is the *opium* of the people.

Karl Marx, "A Contribution to the Critique of Hegel's Philosophy of Right" (1844)

You'll notice two things right off the bat. First, Marx says that religion is the *opium,* not the *opiate,* of the people—a small difference, but worth getting right. (Opium is a particular drug, while opiates are a class.) Second, Marx is very fond of italics.

Characterizing religion as a painkilling drug, shocking as it still is to many, was even more radical in its day. And yet Marx, more than condemning religion itself, was actually critiquing the condition of a society that would lead people to it. Nonetheless, forever after we would hear about "Godless communists," implying that Marxist thought lacks values and morals.

This isn't quite true. What Marx really meant was that religion functions to pacify the oppressed; and oppression is definitely a moral wrong. Religion, he said, reflects what is lacking in society; it is an idealization of what people aspire to but cannot now enjoy. Social conditions in mid-century Europe had reduced workers to little better than slaves; the same conditions produced a religion that promised a better world in the afterlife.

Religion isn't merely a superstition or an illusion. It has a social function: to distract the oppressed from the truth of their oppression. So long as the exploited and downtrodden believe their sufferings will earn them freedom and happiness hereafter, they will think their oppression part of the natural order—a necessary burden rather than something imposed by other men. This, then, is what Marx meant by calling religion the "*opium* of the people": It dulls their pain but at the same time makes them sluggish, clouding their perception of reality and robbing them of the will to change.

What did Marx want? He wanted the "people" to open their eyes to the harsh realities of nineteenth-century bourgeois capitalism. The capitalists were squeezing more and more profit out of the proletariat's labor, at the same time "alienating" workers from self-realization. What workers deserved, and could have if they arose from their slumber, was control over their own labor, possession of the value they created through work, and thus self-esteem, freedom, and power.

To that end, Marx called for the "abolition of religion as the *illusory* happiness of the people." He wanted them to demand "*real* happiness," which in Marx's materialist philosophy was freedom and fulfillment in *this* world. Since the rich and powerful aren't just going to hand these over, the masses shall have to seize them. Thus class struggle and revolution. Would that it were that simple.

Conspicuous Consumption

[The gentleman of leisure can, in sufficiently advanced societies, no longer be] simply the successful, aggressive male,—the man of strength, resource, and intrepidity. In order to avoid stultification he must also cultivate his tastes, for it now becomes incumbent on him to discriminate with some nicety between the noble and ignoble in consumable goods.... Closely related to the requirement that the gentleman must consume freely and of the right kind of goods, there is the requirement that he must know how to consume them in a seemly manner. His life of leisure must be conducted in due form. Hence arise good manners.... High-bred manners and ways of living are items of conformity to the norm of conspicuous leisure and conspicuous consumption.

Thorstein Veblen, *The Theory of the Leisure Class,* chapter 4

Conspicuous consumption is so pervasive now that it's surprising anyone had to *invent* the idea. In fact, it is an ancient one—as ancient as the need to show off or keep up with the Joneses—but the idea had no theorist before American philosopher Thorstein Veblen (1857–1929), coiner of the phrase itself.

Veblen, whose contempt for consumer society is palpable, began by coining the term "leisure class" in his first and most famous book, *The Theory of the Leisure Class* (1899). According to Veblen, as soon as a society has advanced past its primitive stage, and goods are available in surplus of what is required for survival, basic class distinctions will arise. On the one hand are those who spend their days in manual labor; on the other, those who give themselves over to more "honorable" pursuits (chiefly, in the early stages, hunting and warfare). The one class produces, while the other "exploits"—in other words, the one makes, the other takes.

As society develops further, the more elevated class retreats further from any sort of productive labor. This class, through social and political coercion, takes control of communal resources and wealth. Meanwhile, by delegating more and more labor to others, it finds itself with more and more time on its hands. The "leisure class" then devotes this free time to gratifying a very basic human need: the need to be esteemed by others.

"In order to gain and to hold the esteem of men," says Veblen, "it is not sufficient merely to possess wealth or power. The wealth or power must be put in evidence, for esteem is awarded only on evidence." In barbaric societies, the evidence consists of trophies and booty, concrete symbols of successful exploit. In more advanced societies, the evidence is plenty of time and money to waste.

Thus the elevation of such "useless" pursuits as painting, music, fashion, studying dead languages, breeding racehorses, and so forth. All of these require a great deal of time and thus a great deal of freedom from ignoble labor. And thus the conspicuous display of high-cost luxury items, from clothing to cars, carpets to crystal. The less such items have to do with getting a living or with production of any sort, the better.

Showing off one's wealth and power, however, isn't sufficient unto itself. Because "conspicuous consumption," as Veblen called it, is intended to assert one's superiority to others, it is invidious—it generates envy. And thus it generates competition, not only to keep up with the Joneses, but to better them. So displays of wealth and taste are merely bids in the poker game of competitive envy; you must keep upping them to keep up or suffer the humiliation of folding your hand.

None of this would matter much if it were confined to the very richest of the "superior pecuniary class," to use Veblen's term. Unfortunately, the middle class, as soon as one develops, gets drawn into the game, to the detriment not only of themselves but also to the economy. Surplus earnings are channeled not into productive investments, but into displaying one's aspirations. As people are more dominated by ego than by reason (which is one reason classical economics fails), they will often, given the choice, buy a new high-tech toy rather than a savings bond or something actually useful.

Besides training his sarcasm on human vanity, Veblen had much to contribute to social thought. For one, his harshness in *Theory of the Leisure Class* conceals a deep concern for the plight of the working class, which, thanks to laissez faire doctrine, had grown very dire by the turn of the century. His many other, less famous writings pointed the way for the architects of the New Deal to employ social policy to ameliorate the Depression he had predicted. But if his first had also been his only book, we could still thank him for a very entertaining way to pass moments of leisure in reading.

Deficit Spending

English economist and man of letters John Maynard Keynes (1883–1946) was once credited, and is now routinely blamed, for the theory of deficit spending. It is true that Keynes (pronounced "Canes") was a vocal advocate of public works, even unproductive ones. But the fact is that nowhere in his many works do we find the terms "deficit spending" or "deficit financing," or any detailed defense of the welfare state.

What we do find—especially in the book that made his reputation, *The General Theory of Employment, Interest and Money* (1936) —is an ambitious and in its day novel attempt to analyze the interworkings of large-scale or aggregate economic factors. This approach is now known as "macroeconomics," or "The New Economics." At a time when most economists were studying trees, Keynes sized up the forest.

Keynes's essential notion is that we may understand and manipulate national economies if we accept one simple axiom: that national income equals the sum of consumption and investment. Furthermore, the level of national employment is directly proportional to national income. Thus if consumption and investment both grow, so will employment.

Fascinating? No, but the implications are deep. The way Keynes set up his formulas, it turns out that any increase in investment must produce a greater increase in income. In other words, every dollar invested in the economy generates more than a dollar in national income, which in turn gets you more than a dollar's worth of new employment.

Thus Keynes's conclusion: In the normal course of events, public investment—that is, government spending—is always in the best interests of society. This is true even if an income tax might not recoup the expenditure, thus leaving the government in debt—which is where the later coinage "deficit spending" comes from.

Why had no one seen this before? After all, Keynes's arithmetic was pretty elementary. In the first place, economists hadn't been in the habit of looking at the "macroeconomy." In the second place, Keynes made a new assumption about economic behavior. He believed that people are driven by a "propensity to consume" that is relatively fixed with respect to personal income. But the relationship between income and spending is not a direct proportion. Someone who spends $15,000 out of a $20,000 income will not spend $30,000 dollars on double the income. The richer you are, the smaller a percentage of your income you will spend consuming things—the rich don't eat that many more ice-cream cones than the poor.

Just so, the national level of consumption will always lag behind a rise in national income. This is the mathematical basis for Keynes's advocacy of public investment. For if income equals consumption plus investment, and consumption increases more slowly than income, then the ratio between any growth in income and the change in investment that caused it must be greater than one. It is because our consumption habits change more slowly than our income, then, that public investment pays off.

If it also leads to national debt, that's a shame, but Keynes preferred debt to a foolish frugality. Besides, Keynes put no stock at all in the classical economic belief that capitalist economies, left to themselves, will tend toward full employment and a maximum use of resources. He rather thought that unfettered capitalism, by favoring the already wealthy, the greedy, the sly, and the unscrupulous, only leads to an ever-increasing concentration of wealth and to unhealthy levels of unemployment. It is important that Keynes ascended to his peak of influence in the aftermath of the Great Depression, which not only supplied a best-case scenario for the application of his ideas, but also gravely wounded the reputation of laissez-faire.

Especially in such times of depression and low unemployment, private investment tends to shrink, as future returns seem doubtful; consumption also shrinks, and along with both, the national income. Thus the need for public works and other forms of government stimulation of the economy. However, Keynes was also pessimistic about the eventual achievement in Britain or elsewhere of full employment, and he believed that technological stimulation of industry had already peaked in the West.

Keynes seems to have missed the boat regarding technological progress. And many economists tend to doubt one of his most basic assertions: that the relation of consumption to national income is stable, and that the former never grows faster than the latter. In the short run, perhaps not; but in the long run, consumption patterns have demonstrably changed everywhere. On the other hand, as Keynes notoriously quipped, "In the long run we are all dead."

Monetarism

Keynesian macroeconomic theory has long ruled among Western policy makers but there *are* some who find it hard to swallow. As they see it, the problem with Keynes's ideas, at least as they are practiced, is that they produce an unhealthy reliance on fiscal policy—that is, on taxing and spending.

The most influential critic of fiscal policy since the 1960s has been Milton Friedman, a professor at the University of Chicago and the central figure in the "Chicago School" of economics. In the first place, as Friedman noted, if the objective of Keynesian taxing and spending is to promote full employment without boosting inflation, it's been a miserable failure, as any American living through the 1970s could have told you.

In the second place, it is mere folly to suppose that government bureaucrats and policy makers are wiser than the market, whose collective wisdom and autonomous activities ultimately drive economic development. Markets are sufficiently obscure and unpredictable in their actions that mere mortals will usually only make things worse by trying to manipulate them.

Here Friedman embraces the old arguments for laissez faire, and in general his political ideas definitely hark back to the late-eighteenth century. His economic theory—known as "monetarism"—is, however, more novel (though not entirely new). Rejecting Keynes's formulas for investment and consumption, Friedman proposed that income, employment, and prices are far more dependent on the supply of money and the rate it changes hands than on public investment. In his view, the real power of the American economy lies not in the greedy and profligate hands of Congress, but in the sure and steady hands of the Federal Reserve, which controls the money supply and interest rates.

But Friedman, consistent with his political opposition to state control, has never advocated manipulating the money supply in order to "fix" current economic problems. He'd rather leave everything up to the market, with the Federal Reserve acting only in a consistent, stable way to gradually increase the supply of money over time—unwavering and indifferent to the troubles or triumphs of the day.

It should be noted that Friedman isn't trashing Keynes's theories, but rather how they've been used. Keynes himself, who was no fiscal purist, thought the role of banks just as important as the role of government. Every modern economic school, including monetarism, owes a debt to Keynes for establishing basic macroeconomic theory. (Even Richard Nixon, hardly a big tax-and-spender, called himself a Keynesian.) No economist, furthermore, is deep down totally devoted to laissez faire—if he were, he wouldn't have any reason to study economics. Indeed, it's hard to see what Friedman does all day.

Parkinson's Law

"Work expands so as to fill the time available for its completion."

C. Northcote Parkinson, *Parkinson's Law, or the Pursuit of Progress* (1957)

You've probably noticed that if you have ten minutes to write a letter, you will do it in ten minutes; but if you have four hours, you'll take four hours. Such is the gist of "Parkinson's Law," first revealed in 1955 by the historian Cyril Northcote Parkinson in the pages of *The Economist*. Parodying the typical essay in socio-economics, Parkinson "proved" his claim by charting growth in the British naval bureaucracy at a time when its responsibilities were diminishing: It was taking more people to accomplish less.

"It is the busiest man who has time to spare," quoth Parkinson. People tend to make work for themselves; what varies is not free time but efficiency. Chiefly interested in how his law applied to the workplace, Parkinson wryly observed that "An official wants to multiply subordinates, not rivals" and that "Officials make work for each other." Regardless of the amount of real work, managers continue to hire more inferiors, just to make themselves seem more responsible and powerful—which sets off a chain reaction requiring more underlings and more supervision, without any appreciable increase in productivity.

Despite Parkinson's irony, his law rings true both at the office and at home. The busier you are, the more efficient you need to be. The emptier your day, the more demanding simple tasks become. Given human nature, unfinishable tasks—such as spring cleaning—are something of a godsend.

Before We Leave You, A Few More Thoughts

Luddism

"Luddism" by its current meaning—"the fear of technology and a resistance to progress"—is a pretty sorry excuse for a great idea. But the historical Luddites offer a more complex and interesting story; and while their day in the sun was brief (1811–1816), their concerns still count, especially in developing countries.

The movement allegedly took its name from the Leicestershire malcontent Ned Ludd, whom the Oxford English Dictionary refers to as a "lunatic living about 1779." Nobody really knows exactly what Ludd perpetrated, but one story is that he destroyed a stocking-knitting machine at the factory where he worked because the boss had called him on the carpet. Whatever the truth, Ludd became a countryside folk hero, and the pseudonyms "King Ludd" and "General Ludd" were adopted by certain anti-industrial rebels.

Unlike Ludd, the Luddites of the early nineteenth century weren't factory workers, or at least didn't start out that way. They

Eureka!

were mostly highly skilled village craftsmen, principally in Yorkshire, practicing various trades in the wool industry. They were well paid, and what's more, they were community leaders with strong family bonds, a devotion to domestic industry, and a dedication to tradition.

But things were changing fast. Hurting from the American Revolution and the Napoleonic Wars, the English economy was shrinking. At the same time, industrialism was rapidly gaining momentum as factories and automated machinery spread through the countryside. Men, women, and children were forced into factories, and traditional ways had to yield to the discipline of the workplace. To the craftsmen this meant both the destruction of their small businesses and the rending of the fabrics of home and community life.

The American Revolution was recent history, and it proved inspiring to the folk of Yorkshire, who had been abandoned by their politicians. In a highly organized and systematic fashion, the craftsmen and their allies—the "Luddites"—broke into mills at night and smashed weaving frames to bits. It was sort of like the Boston Tea Party with hammers.

Righteous rebellion or industrial terrorism? History belongs to the victors, and the losers paid dearly. Many were shot or hanged, and now the word *Luddite* is almost interchangeable with *kook*. The understandable rage of the Yorkshire craftsmen against the capitalists (not against machines) is currently equated with the quixotic refusal to buy an answering machine on grounds of principle. Of course, today few believe that the march of technology can be halted, but at the time the outcome wasn't a foregone conclusion. And there are still countries, such as India, where a tradition of village craft survives and is likewise imperiled by progress. True Luddism is alive in such places, but history isn't encouraging.

The Pathetic Fallacy

[There is] a fallacy caused by an excited state of the feelings, making us, for the time, more or less irrational.... All violent feelings have the same effect. They produce in us a falseness in all our impressions of external things, which I would generally characterize as the "Pathetic fallacy."

John Ruskin, "Of the Pathetic Fallacy," in
Modern Painters (1856), Part IV, chapter 12

Does your stomach complain about pepperoni pizza? Does money burn a hole in your pocket? Do the heavens smile on your latest romance? If so, you're guilty of the "pathetic fallacy," but don't take it personally: Most everyone is.

Anyway, *pathetic* doesn't mean what you think it does. The Victorian critic John Ruskin (1819-1900), coiner of the phrase, had in mind the root meaning of *pathos,* which is Greek for "emotion." The fallacy he meant was the kind of mistake made principally by creative types—namely, granting feelings, intentions, and other human traits to objects that can't have them. In short, the pathetic fallacy is a species of anthropomorphism.

This sort of mistake is not intellectual. We don't *really* think a stomach can complain, or that the sea can be angry, or that that chair intended to bruise our toe. But sometimes, driven by passion (what Ruskin calls "violent feeling"), we do see things that way. In the grip of ecstasy or anger, our perceptions are colored, and our imagination conquers our reason.

Which isn't necessarily a bad thing. In fact, for poets and painters it is the stuff of their trade. Depicting a struggle to shore, poet Alton Locke wrote, "They rowed her in across the rolling foam— / The cruel, crawling foam." Of course, Ruskin notes, "foam is not cruel, neither does it crawl." In another case, Oliver Wendell Holmes wrote fancifully of "The spendthrift crocus, bursting through the mould / Naked and shivering, with his cup of gold." This, Ruskin points out dryly, "is very beautiful, and

Eureka!

yet very untrue. The crocus is not a spendthrift, but a hardy plant; its yellow is not gold, but saffron." The claims of these poets are thus fallacious, but partly for that very reason we delight in the poetry.

But emotional effects and anthropomorphism do not suffice to make good poetry. Neither are they necessary since, as Ruskin points out, pathetic fallacy is difficult to find in the works of Shakespeare, Homer, or Dante. (It is, however, easily found in "reflective" poetry, especially that of the Romantics—Coleridge, Wordsworth, Shelley, Keats, and the rest.) Furthermore, it isn't really "untruth" in itself that pleases us; it is rather, to put it paradoxically, *truthful* untruth. If a poet is going to commit the fallacy, she must ensure that the confusion or untruth is compatible with the emotion supposedly inspiring it. It does not do, in fact it's displeasing, to depict a man gripped by anger admiring the smile of the sun.

In short, the effects of feeling should match the strength and character of feeling. Among the greatest poets, such as Shakespeare, strength of feeling is matched by strength of thought, so that fallacy is avoided: Feelings may be *associated* with a primrose, but the primrose is still just a primrose. Likewise, there is no fallacy committed by those who feel little or nothing, for they just see things as they "really" are. The fallacy arises either when a poet of some talent but not great intellect is overcome with emotion (Romantic poetry being one result) or when a strong, thinking poet is subjected to an unusually intense experience, glimpsing some transcendent, overwhelming truth. This is inspired fallacy, and it is the best sort.

On the other hand, parroted metaphors, inauthentic emotion, and poetic pretension—the kind of derivative and overwrought stuff one writes after a breakup—are just bad and unforgivable. To the degree the pathetic fallacy is fanciful rather than felt, it is merely fallacy; to the degree it is emotionally true, it is pathetic. And so, Ruskin concludes, "the pathetic fallacy is powerful only in so far as it is pathetic, feeble so far as it is fallacious."

Ruskin's idea would later reemerge in the jargon of psychoanalysis. What he called "the pathetic fallacy" Freud called "projection," the transfer of subjective sensations and emotions to objects in the outside world. "The projection of inner perceptions to the outside," Freud wrote, "is a primitive mechanism which, for instance, also influences our sense perceptions, so that it normally has the greatest share [among psychic defenses] in shaping our world." And so, according to J. A. C. Brown, "whenever the internal and subjective becomes confused with the external and objective we may speak of projection." Not only do poets and artists project; so do we all.

"Form Follows Function"

All things in nature have a shape, that is to say, a form, an outward semblance, that tells us what they are, that distinguishes them from ourselves and from each other....

Whether it be the sweeping eagle in his flight or the open apple-blossom, the toiling work-horse, the blithe swan, the branching oak, the winding stream at its base, the drifting clouds, over all the coursing sun, form ever follows function, and this is the law. Where function does not change form does not change. The granite rocks, the ever-brooding hills, remain for ages; the lightning lives, comes into shape, and dies in a twinkling.

Louis Sullivan, "The Tall Office Building Artistically Considered" (1896)

Louis Sullivan (1856–1924) thought American architecture at the end of the century to be in a mighty sorry state. Dynamic America, burgeoning economic power, home of the first skyscrapers, was erecting buildings out of the past—buildings that invoked Greece, Rome, the Gothic and the Baroque, the Renaissance and the Enlightenment—everything but contemporary America.

Professional architecture was at that time largely an historical enterprise, highly artificial, focused on tradition and given to arbitrary ornamentation. The emphasis lay on the architect's invention and his scholarship. Sullivan, however, insisted that architects should construct buildings on *natural* rather than artificial principles. And the most important determinant in natural architecture is what a building is *for*—its purpose, its essence, its *raison d'être*. He called this the building's "function" and put forth in 1896 his famous dictum, now better known than its author: "Form ever follows function." (That "ever," which so spoils the alliteration, is ever dropped.)

Sullivan coined the phrase in an essay on what was then called the "tall office building." (We call it a "skyscraper"; the first of these, the Home Insurance Building in Chicago, had been built thirteen years before Sullivan's essay.) What is the essence of such a building, Sullivan asked? What is its *function*? For the person on the street, the building should communicate loftiness, exaltation, and ambition. The natural architectural principle that follows is to design the building so that its rise is emphasized and uninterrupted. But the way most skyscrapers were being built, ornamentation and whimsical variations broke the line of rise, interrupting the movement of eye from base to peak.

So one of the building's functions is already betrayed by its form. There are other functions, looked at from the perspective of the building's use rather than its aesthetic effect. The ground floor and second story have their purpose: to house businesses and banks, to provide an open and inviting space for visitors and workers, to furnish liberal lighting and freedom of access. Such purposes may be naturally served by particular architectural forms: "liberal, expansive, [and] sumptuous," "based exactly on the practical necessities, but expressed with a sentiment of largeness and freedom."

But what of the rest of the building? Every floor between the second and the topmost will be functionally identical: each "tier just like another tier, one office just like all the other offices." This is the practical requirement of efficient use of space. In nature, Sullivan points out, things that serve the same function (say, flight) all have the same form (that is, wings). So long as a natural object continues to serve the same purpose, it maintains the same form. By "function" Sullivan means something like "natural essence": The bird's form expresses the fact and essence of being a bird, as opposed to something else; there is no bird that looks like a monkey, no rock that looks like a tree. (This is sort of a tautology.) So a bank should not look like a Greek temple or Gothic manor; so every story of a building that serves the same function should have the same form.

This may seem obvious today, as the skyscrapers we see virtually all observe Sullivan's dictum. But such was not the case at the time. Sullivan condemned the sixteen-story building consisting of "sixteen separate, distinct and unrelated buildings piled one upon the other until the top of the pile is reached." Such monstrosities were not designed by ignorant or naïve architects, but "trained" ones beset by fears of seeming unsophisticated or unresourceful.

Sullivan overstates the case a bit. And in fact he grew more bitter with time, as his business declined, negatively affected by his radical social agenda. But while his practice lasted he did begin reversing the conventions of his day. Working from a philosophy of natural democracy and organic growth, Sullivan attempted to bring the spirit of nature to the architecture of American commerce. (Perhaps the best surviving example is his Wainwright Building in St. Louis, built 1890–1891.) His ideas were not completely original, but they were influential, especially when put into practice by Sullivan's protégé Frank Lloyd Wright, the champion of "organic architecture."

Between them, Sullivan and Wright brought Modern architecture to America, with its emphasis on space and bare structure, functionality and modernity. Taken to the extreme of naked utility, such principles were capable of producing ugly and alienating buildings, such as the worst examples of the "International Style." But the blame cannot really be laid on Sullivan or on his often misunderstood catchphrase "Form ever follows function." He was no enemy of aesthetic values, or even of decoration and ornament, which he only wished to render organic rather than artificial. He called for an architecture that responds to spiritual as well as utilitarian needs and that expresses the spirit of the age.

"And thus," he proclaims in the essay's peroration, "when native instinct and sensibility shall govern the exercise of our beloved art; when the known law, the respected law, shall be that form ever follows function," then "it may be proclaimed that we are on the high-road to a natural and satisfying art, an architecture that will soon become a fine art in the true, the best sense of the word, an art that will live because it will be of the people, for the people, and by the people." But not, of course, *entirely* original.

"Less Is More"

The great paradoxes of Western thought, it must be said, have rarely had great practical consequences. Chances are good that if you start your car, point it at the mall, and go, you're going to get there sooner or later, no matter what Zeno said [see p. 21]. Likewise, Russell's paradox [p. 108] may have wreaked havoc with set theory, but in the aftermath life pretty much went on as before.

The exception that proves the rule is "Less is more," a favorite catchphrase of German architect Ludwig Mies van der Rohe (1886–1969). Actually, Mies didn't coin the phrase; neither did he invent "God is in the details," another slogan credited to him. "Less is more" had already appeared in Robert Browning's great poem, "Andrea del Sarto" (1855) and was floating around the German art world in various forms. (Mies' most direct source was his teacher, Peter Behrens.) But it was Mies who made a practice of the paradox, to the delight of some and, these days, to the derision of many.

What he meant by the phrase is this: A building should be constructed on its essentials; extra trappings or additions only distract from clarity, utility, and effect. (This is essentially how we use the phrase today: "More of a good thing isn't necessarily better.") What the architect seeks is not "less" for its own sake—merely stripping a building down to its structure—but rather what is appropriate to the materials, location, and required design.

This goal superficially resembles that of Louis Sullivan, who championed organic unity with his phrase "Form ever follows function." Mies, however, was more focused on rationality and precision than the metaphysical Sullivan. The immediate "function" of a building did not overly concern him; he, unlike Sullivan, foresaw that any structure might in the future be put to a variety of uses, serve a variety of functions. This is one rationale of the quest for simplicity: The more open and pure the building, the more adaptable it would be.

In practice, Mies' dictum "Less is more" resulted in buildings of geometric rigidity emphasizing, rather than attempting to hide, their materials of construction. The most famous example, at least in America, is the Seagram Building on Park Avenue in New York, which Mies designed in the late 1950s with Philip Johnson. Extremely regular in construction, the Segram is structurally almost bare, a rigid tower of glass and bronze. Though not his first, it is *the* prototypical "glass box," which inspired endless inferior imitations in subsequent decades. Less of this style would definitely be more.

"Those Who Cannot Remember the Past Are Condemned to Repeat It"

Progress, far from consisting in change, depends on retentiveness. When change is absolute there remains no being to improve and no direction is set for possible improvement: and when experience is not retained, as among savages, infancy is perpetual. Those who cannot remember the past are condemned to repeat it.

George Santayana, *The Life of Reason* (1905), Volume I, chapter XII

In becoming a cliché, this observation by the Spanish-American philosopher George Santayana (1863–1952) has lost all its depth. Usually heard in the form "Those who *do not* remember the past...," it has been reduced to advice on a proper curriculum. "Learn your history, boys and girls, or the next time [insert atrocity here] comes along, you won't remember what happened the first time."

Not that this is false; it's just not what Santayana meant. He chose the word *cannot* for a reason—namely, because he meant "are literally unable to." Such is the fate of infants and "savages," for whom every day dawns anew, the experiences and lessons of yesterday having been forgotten. It is not that such people (one could argue with the term "savages") choose to be ignorant; it's that they are incapable of historical thought.

In this condition of forgetfulness, a person is unable to make any informed decisions or to advance himself. He will simply continue to act according to instinct and reflex, which are by their nature repetitive. Every day is more or less the same day, which is what Santayana means by "repeating the past."

Santayana's larger point is that progress requires a certain stability and "retentiveness" in individuals and societies. This is the basis for human evolution, which is modeled on Darwin's evolution of species: Educated behavior, based on experience, is more likely to succeed in the face of changing conditions. That is, we will get better and better at dealing with the ever-changing world if we are both "retentive" and "flexible": conscious of the past and yet adaptable.

The larger context for Santayana's speculations is his "naturalism," better known as "materialism." Man, in his view, is wholly and completely the product of nature; and the mind is nothing more than the natural activity of the brain. Given that nature is constantly in flux, so too is what we call "human nature." The beliefs, values, thought processes, instincts, and desires of the ancient Greeks are very different from those of medieval Europeans or contemporary Africans.

There is no such thing, therefore, as a "universal law," if by that we mean rules applicable across time and space. Yet at the same time, in any *particular* time and place, men and women do share beliefs, values, thought-processes, and the rest. Otherwise there could be no communication at all. And such a *particular* human nature has a potential "ideal state," in which it is everything it can be: ideally suited, within its limits, to the time and the conditions. Each individual has his or her own ideal, which has nothing to do with what the majority of people think, feel, or do.

In fact, Santayana believed deeply that people are unequally graced with reason and talents. It might just be the ideal of some to work on assembly lines, while it is the ideal of others to run the state. He wasn't, therefore, an enthusiast of democracy. In his view Nature herself is undemocratic; some species die off while others flourish and evolve, and this is because some species are superior to others. In men and women, sharpness of reason and remembrance of the past are suited to progress and self-realization, to achieving one's ideal. So hit those history books now.

APPENDIXES

Equations

Einstein's Equation of Energy and Mass (page 98)

Here's the basic idea: As an object of mass m (which is different from its weight, which depends on gravity) traveling with a constant velocity v, has a momentum which may be expressed as the product mv. The object's inertia, which is proportional to its mass, m, will keep it moving in the same direction at the same speed unless force is applied to it, causing an acceleration (a change in v). This added force adds energy to the object, which under normal circumstances is expressed as an increase in velocity (extra momentum). The total energy expressed in the object's motion is called its "kinetic energy," which is its momentum times its velocity divided by two—that is,

$$E = m\frac{v^2}{2}$$

This gives the amount of energy required to set a body of mass m at rest into motion of velocity v.

This equation holds true, however, only under Newtonian relativity. If we take into account the Lorentz equations of the special theory (which describe contraction of space and dilation of time in the direction of relative motion), the formula becomes:

$$E = \frac{mc^2}{\sqrt{1 - \dfrac{v^2}{c^2}}}$$

where c is the speed of light in a vacuum. As the object's velocity, v, approaches that of the speed of light, v^2/c^2 approaches 1, and the denominator disappears. In this case, because we're dividing by zero, E is infinite: in other words, it would take an infinite amount of energy to propel an object to the speed of light. In short, no object with any mass at all could ever travel at the speed of light—there's not enough energy to get it there. (Light itself has no mass, by the way.)

Now let's look at the opposite case: Energy is added to an object of mass m in motion at velocity v, but we somehow prevent the object from moving any faster than v. For convenience (and it would work out the same otherwise), let's assume that the object was at rest to begin with—it had no kinetic energy at all—and that v remains 0 despite the addition of the energy E. If $v=0$, so does v^2/c^2, and the denominator of our equation reduces to 1. In which case,

$$E = mc^2$$

In other words, the energy added must have all become mass, since mass is the only variable left that can change. The added mass, if we perform elementary division, will be equal to E/c^2.

But mass is just mass, whether or not it is produced by an addition of energy. Therefore Einstein's formula holds in every case; if we wish to know how much energy is latent in a body of mass 10 grams, we simply multiply 10 grams by the speed of light squared, and convert to the appropriate units. Get out your calculators!

Chaos and the "Logistic Equation" (page 143)

The logistic equation, which played a crucial role in the development of chaos theory, is a variant of a simple linear equation. Suppose we're studying population growth among a particular group of animals—say, the gray squirrels in Central Park. Our first hypothesis is that the squirrel population grows at a steady rate from year to year, say .1 or 10%. In that case, the population in the year $n+1$ will be 1.1 (100% + 10%) times the population in year n, or, to put it in mathematical form, $x_{n+1} = 1.1(x_n)$, where x_n is the population in year n. The rate of change (1.1) is fixed.

But the more observing we do, the more we realize that population growth is not steady at all, but that the rate of change itself changes along with the size of the population. Instead of simply applying a fixed multiplier from year to year, we are forced to introduce a nonlinear factor. A better predictor of the squirrel population in Central Park is the nonlinear logistic equation, which looks like this:

$$x_{n+1} = r x_n (1 - x_n)$$

where x_n stands for the population in year n, expressed as a percent of maximum total population, and where r stands for some fixed factor of change. (If the maximum squirrel population in Central Park is 1,500, and in a given year n the actual population is 1,000, then $x_n = 1,000/1,500 = .667$.)

The logistic equation resembles our original linear one save for the additional (nonlinear) factor $1 - x_n$, which grows smaller as the population increases and grows larger as the population decreases. (Since x_n is a percentage—that is, a number between 0 and 1—then $1 - x_n$ will always be positive; thus the population never sinks below zero.) The equation, furthermore, is called "iterative," because the results from one year are plugged back in to get the results of the next; that is, the equation is a "feedback loop."

As it turns out, the logistic equation, which is nonlinear (its rate of change is variable), has a number of very interesting properties. For certain values of r (namely, values less than 3), the population will eventually hone in on one precise, unchanging quantity. It doesn't even matter what number you plug in first, so long as it's small but not equal to zero. This destination—the value x_n approaches as you keep repeating the equation—is called an "attractor." Even more interesting, as r exceeds 3, the population eventually hones in on two alternating values, getting near to one in any given year and then near to the other in the next year. The attractor has split and is now called an "attractor of period 2." Furthermore, if we increase r to approximately 3.45, the attractor splits into four, and later into eight, and later into sixteen, and so on and on. But it doesn't keep doubling forever; at a certain point, when r is approximately equal to 3.57, the attractor becomes unpredictable, wildly variant, seemingly totally chaotic. (At this point it's called a "strange attractor.") But as it turns out, this is chaos with a pattern.

The logistic equation isn't the only one to produce splitting attractors and patterned chaos. As Mitchell Feigenbaum discovered in the 1970s, there are all sorts of equations, many of them commonly used by practical scientists, that when rigged to a feedback loop look from the results exactly like the logistic equation (one example is $r \sin \pi x$). Mathematicians realized this couldn't be just an odd coincidence, and Feigenbaum's discovery really got the chaos ball rolling.

Sources

Aristotle, *The Basic Works of Aristotle,* ed. Richard McKeon (New York: Random House, 1941).

Saint Augustine, *On the Two Cities: Selections from 'The City of God,'* trans. Marcus Dods (New York: Frederick Ungar, 1957).

Francis Bacon, *Essays, Advancement of Learning, New Atlantis, and Other Pieces,* ed. Richard Foster Jones (New York: Odyssey, 1937).

Jacques Derrida, *Of Grammatology* (1967), trans. Gayatri Chakravorty Spivak (Baltimore: Johns Hopkins University Press, 1976).

———, *Positions* (1972), trans. Alan Bass (Chicago: University of Chicago Press, 1981).

Albert Einstein, *Relativity: The Special and General Theory,* trans. Robert W. Lawson (New York: Wings Books, 1961).

———, *Beyond the Pleasure Principle* (1920), trans. James Strachey (New York: Norton, 1961).

———, *The Ego and the Id* (1923), trans. Joan Riviere and James Strachey (New York: Norton, 1962).

———, *General Psychological Theory,* ed. Philip Rieff (New York: Macmillan, 1963).

———, *New Introductory Lectures on Psychoanalysis* (1933), trans. James Strachey (New York: Norton, 1965).

———, *Sexuality and the Psychology of Love,* ed. Philip Rieff (New York: Macmillan, 1963).

G.W.F. Hegel, *Phenomenology of Spirit,* trans. A. V. Miller (Oxford: Clarendon, 1977).

The Holy Bible, King James Version (1611).

Edmund Husserl, *Ideas: General Introduction to Pure Phenomenology,* trans. W. R. Boyce Gibson (New York: Collier, 1962).

C. G. Jung, "On the Psychology of the Unconscious," trans. R.F. C. Hull, excerpted in *Great Ideas in Psychology,* ed. Robert W. Marks (New York: Bantam, 1966).

Melanie Klein, *Love, Guilt and Reparation & Other Works 1921–1945* (New York: Dell, 1975).

Steven Knapp and Walter Benn Michaels, "Against Theory," *Critical Inquiry* 8 (Summer 1982).

Thomas S. Kuhn, *The Structure of Scientific Revolutions,* second edition (Chicago: University of Chicago Press, 1970).

Robert W. Marks, ed., *Space, Time, and the New Mathematics* (New York: Bantam, 1964).

Karl Marx, *Early Writings,* ed. Quintin Hoare (New York: Vintage, 1975).

Marshall McLuhan, *The Gutenberg Galaxy: The Making of Typographic Man* (Toronto: University of Toronto Press, 1962).

——, *Understanding Media: The Extensions of Man,* revised edition (New York: McGraw Hill, 1965).

—— and Edmund Carpenter, eds., *Explorations in Communication* (Boston: Beacon Press, 1960).

Friedrich Nietzsche, *The Portable Nietzsche,* trans. and ed. Walter Kaufmann (New York: Viking, 1968).

Plato, *The Collected Dialogues of Plato,* ed. Edith Hamilton and Huntington Cairns, corrected edition (Princeton: Princeton University Press, 1963).

Plutarch, "The Impossibility of Pleasure according to Epicurus," trans. William Baxter (New York: n.p., 1859)

Karl R. Popper, *The Logic of Scientific Discovery,* second edition (New York: Harper & Row, 1968).

Jean-Jacques Rousseau, *The Social Contract, or Principles of Political Right,* trans. Henry J. Tozer (London: George Allen & Unwin, 1895).

John Ruskin, *Modern Painters,* 5 vols. (London: Dent, 1906).

Jean-Paul Sartre, *Being and Nothingness* (1943), trans. Hazel E. Barnes (New York: Philosophical Library, 1956).

Louis Sullivan, *The Public Papers,* ed. Robert Twombly (Chicago: University of Chicago Press, 1988).

Philip Wheelwright, ed., *The Presocratics* (New York: Odyssey Press, 1966).

Norbert Wiener, *The Human Use of Human Beings: Cybernetics and Society,* revised edition (Garden City, N.Y.: Anchor/Doubleday, 1954).

References

Mortimer J. Adler, *The Great Ideas: A Lexicon of Western Thought* (New York: Macmillan, 1992).

S. T. Bindoff, *Tudor England* (Harmondsworth: Penguin, 1950)

Werner Blaser, *Mies van der Rohe: Less Is More* (New York: Waser Verlag Zürich, 1986).

I. M. Bochenski, *Contemporary European Philosophy,* trans. Donald Nicholl and Karl Aschenbrenner (Berkeley: University of California Press, 1961).

J. A. C. Brown, *Freud and the Post-Freudians* (Harmondsworth: Penguin, 1961).

Douglas Bush, *English Literature in the Earlier Seventeenth Century 1600–1660* (New York: Oxford University Press, 1945)

Jeremy Campbell, *Grammatical Man: Information, Entropy, Language, and Life* (New York: Simon & Schuster, 1982).

Richard and Fernande DeGeorge, eds., *The Structuralists: From Marx to Lévi-Strauss* (Garden City, N.Y.: Anchor/Doubleday, 1972).

Dictionary of the History of Ideas: Studies of Selected Pivotal Ideas, ed. Philip P. Wiener, 5 vols. (New York: Scribner's, 1973).

Timothy Ferris, *Coming of Age in the Milky Way* (New York: William Morrow, 1988).

James Gleick, *Chaos: Making a New Science* (New York: Viking, 1987).

Stephen Jay Gould, *Ontogeny and Phylogeny* (Cambridge, Mass.: Harvard University Press, 1977).

Stephen W. Hawking, *A Brief History of Time: From the Big Bang to Black Holes* (New York: Bantam, 1988).

Douglas R. Hofstadter, *Metamagical Themas: Questing for the Essence of Mind and Pattern* (New York: Basic Books, 1985).

Judy Jones and William Wilson, *An Incomplete Education* (New York: Ballantine, 1987).

Joseph J. Kockelmans, ed., *Phenomenology: The Philosophy of Edmund Husserl and Its Interpretation* (Garden City, N.Y.: Anchor/Doubleday, 1967).

Bart Kosko, *Fuzzy Thinking: The New Science of Fuzzy Logic* (New York: Hyperion, 1993).

J. Laplanche and J.-B. Pontalis, *The Language of Psychoanalysis,* trans. Donald Nicholson-Smith (New York: Norton, 1973).

Frank Lentricchia, *After the New Criticism* (Chicago: University of Chicago Press, 1980).

Robert H. March, *Physics for Poets,* second edition (Chicago: Contemporary Books, 1978).

A.E.E. McKenzie, *The Major Achievements of Science,* vol. 1 (Cambridge: Cambridge University Press, 1960).

Vincente Medina, *Social Contract Theories: Political Obligation or Anarchy?* (Savage, Md.: Rowman & Littlefield, 1990).

Louis Menand, "An American Prodigy," review of *Charles Sanders Peirce: A Life,* by Joseph Brent, *The New York Review of Books,* December 2, 1993.

Gerald Messadié, *Great Scientific Discoveries,* trans. Alison Twaddle (Edinburgh: Chambers, 1991).

Hugh Morrison, *Louis Sullivan: Prophet of Modern Architecture* (New York: Norton, 1935).

Ernest Nagel and James R. Newman, *Gödel's Proof* (New York: New York University Press, 1958).

Jacob Oser, *The Evolution of Economic Thought,* second edition (New York: Harcourt, Brace & World, 1970)

Donald Palmer, *Looking at Philosophy: The Unbearable Heaviness of Philosophy Made Lighter* (Mountain View, Calif.: Mayfield, 1988).

William Poundstone, *Prisoner's Dilemma* (New York: Doubleday, 1992).

Alex Preminger, ed., *Princeton Encyclopedia of Poetry and Poetics,* enlarged edition (Princeton, N.J.: Princeton University Press, 1974).

Howard Rheingold, *Virtual Reality* (New York: Simon & Schuster, 1991).

Eric Roll, *A History of Economic Thought,* fourth edition (London: Faber and Faber, 1973)

Leland M. Roth, *A Concise History of American Architecture* (New York: Harper & Row, 1979).

James Trefil, *1001 Things Everyone Should Know about Science* (New York: Doubleday, 1992).

Various contributors, "Who Were the Luddites, Really?" History Conference, Topic 139 (October 1992–present), the Whole Earth 'Lectronic Link computer conferencing system, well.sf.ca.us.

W.P.D. Wightman, *The Growth of Scientific Ideas* (New Haven: Yale University Press, 1953).

INDEX

Achilles 21
Aeschylus 170
aether 6, 79, 81
anima 179
animus 178
Anselm of Canterbury 11–14
Aquinas, St. Thomas 7–9
archetypes 178–79
Archimedes 77–78
Archimedes' principle 77–78
Aristarchus of Samos 79
Aristotle 5–8, 22, 25–26, 42, 60, 79–81, 84, 114, 124, 146–47
artificial intelligence 116, 136
attractors 241–42
Augustine of Hippo, St. 1–3
Bacon, Francis 33, 82–83
"Bad money drives out good" 204
bad objects 180–81
Balzac, Honoré de 186
Barthes, Roland 186
base 213
Bateson, William 154
Behaviorism 159–62
Behrens, Peter 235
Bentham, Jeremy 54–55
Big bang 137–41
biogenesis 146
Bohr, Niels 100–2
Boltzmann, Ludwig 132
Born, Max 105

bourgeoisie 213
Brahe, Tycho 80–81
Brown, J.A.C. 230
Browning, Robert 235
Buffon, Georges 150
buoyancy 77
"Butterfly Effect" 142
Campbell, Joseph 179
capitalism 213–14, 216
Carnot, Nicolas Léonard Sadi 129–32
Cartesian coordinate system 27, 97
Categorical imperative 40–41
causality 6–7, 32–35
chaos theory 142–45, 240–42
Charles I, King 42
Chicago School 223
"chicken" 122–23
Chomsky, Noam 188–90
chromosomes 155–56
class struggle 213–16
Clausius, Rudolf Julius Emanuel 129–30
Cogito ergo sum 27–29
Coleridge, Samuel Taylor 229
collective unconscious 178–79
Columbus, Christopher 79
communism 215–16
condensation 164–65
conditioned reflex 158
conspicuous consumption 217–19

Eureka!